Methods of

Intermediate Problems for Eigenvalues

Theory and Ramifications

This is Volume 89 in
MATHEMATICS IN SCIENCE AND ENGINEERING
A series of monographs and textbooks
Edited by RICHARD BELLMAN, *University of Southern California*

The complete listing of the books in this series is available from the Publisher upon request.

Methods of Intermediate Problems for Eigenvalues

THEORY AND RAMIFICATIONS

Alexander Weinstein

Georgetown University
Washington, D. C.

AND

William Stenger

Ambassador College
Pasadena, California

1972

ACADEMIC PRESS New York and London

ACADEMIC PRESS, INC.
111 Fifth Avenue, New York, New York 10003

United Kingdom Edition published by
ACADEMIC PRESS, INC. (LONDON) LTD.
24/28 Oval Road, London NW1 7DD

LIBRARY OF CONGRESS CATALOG CARD NUMBER: 70-154367

AMS(MOS) 1970 Subject Classification 49G15

PRINTED IN THE UNITED STATES OF AMERICA

Contents

CHAPTER THREE

The Classical Maximum-Minimum Theory and Its Extension to Unbounded Operators

CHAPTER FOUR

Intermediate Problems of The First Type

CHAPTER FIVE

Intermediate Problems of The Second Type

CHAPTER SIX

Various Other Methods and Their Connections with Intermediate Problems

CHAPTER SEVEN

The New Maximum-Minimum Theory

CHAPTER EIGHT

Inequalities for Eigenvalues of Parts and Projections of Operators

CHAPTER NINE

Intermediate Problems and Perturbation Theory

Preface

The purpose of this book is to give a detailed exposition of two complementary variational methods for the computation of upper and lower bounds for eigenvalues, namely the Poincaré–Rayleigh–Ritz method and the method of intermediate problems and its important variants and ramifications. Since the appearance in 1966 of the second edition of S. H. Gould's book, *Variational Methods for Eigenvalue Problems: An Introduction to the Weinstein Method of Intermediate Problems*, there have been several very significant developments in the theory of intermediate problems of both a numerical and theoretical nature which made it desirable to present a systematic treatment of the whole subject.

The contributions of intermediate problems to variational calculus, functional analysis, operator theory, and perturbation theory should be of interest to pure mathematicians. On the other hand, engineers, physicists, and applied mathematicians will find that intermediate problems have yielded high precision numerical results for a great variety of problems of classical and quantum mechanics. Actually Appendix B contains tables for fifteen such problems. None of these problems is an artificial example which would not convince the reader of the applicability of the method.

A detailed description of the contents of the various chapters is given in the Introduction.

Acknowledgments

Our gratitude first of all goes to Professor Richard Bellman who took the initiative and encouraged us to write this monograph.

We wish to thank the Air Force Office of Scientific Research which originally partially supported our work on this monograph under grant AFOSR 1306-67 at the American University. Our particular gratitude goes to the National Science Foundation whose grant GP-8903 at Georgetown University made possible the completion of this monograph. We are grateful to Georgetown University for providing a pleasant and stimulating environment which encouraged our work.

We would also like to thank Miss Donnajo Soviero of American University for typing some of the rough drafts of the manuscript. Finally, we wish to thank Ambassador College for helping us to prepare the book for publication, in particular Mrs. Rose Gerke and Mrs. Lyla Yelk for typing the manuscript, Mr. Herbert Vierra for drawing the figures, and Mr. Paul Guy for his assistance in editing the book.

Introduction

This book deals with a class of eigenvalue problems that originated in classical and quantum mechanics. Since the eigenvalues are not explicitly known in most cases, several methods for their approximation have been introduced and developed over a period of years. The long history of these problems is more involved than is usually assumed and some important relationships among the various ideas will be discussed.

While the exposition here will be, as is usually done now, for operators in a Hilbert space, let us emphasize that all of these theories originated in problems of classical physics, mechanics, and quantum theory. These connections with concrete problems proved to be most fruitful. Weber, Poincaré, and Weyl considered the vibration of a membrane. Besides the membrane, Weyl also investigated the classical problem of the electromagnetic cavity. Rayleigh studied organ pipes and Ritz studied the vibration of plates. Weinstein introduced the method of intermediate problems by linking the buckling and vibration of a plate to the vibration of a membrane. The fact that all of these theories started on special cases is hardly ever mentioned in the modern literature except perhaps in the case of intermediate problems, which alone has been singled out as a theory which began with a special case. Even recently, after intermediate problems were reformulated in Hilbert space by Weinstein and Aronszajn and extensively investigated there by Aronszajn, the recent important advances by Bazley and Fox were motivated by concrete problems of quantum theory. The theory of intermediate problems was used in a new way in a method introduced by Fichera. Various other significant numerical and theoretical results have been given by Börsch–Supan, Colautti, Diaz,

Fujita, Gay, Howland, Kato, Kuroda, Löwdin, Metcalf, Payne, Rubinstein, Sigillito, Stadter, Stakgold, Stenger, Velte, Weinberger, and others.

Our whole discussion will be dominated by the variational approach. This will lead to inequalities that can be interpreted as upper and lower bounds for eigenvalues, thereby determining the accuracy of the methods. Since there is no method that gives a sharp estimate of the error in an approximation, the only reliable way is to use two complementary methods that give upper and lower bounds respectively.

The oldest method for obtaining numerical upper bounds is the variational method originally known as *Rayleigh's principle*. In fact, it is obvious that the very definition of the lowest eigenvalue as the minimum of the Rayleigh quotient provides a means of obtaining an upper bound. A further development for higher eigenvalues usually, but not quite correctly, called the Rayleigh–Ritz method will be given in Chapter 2.

A much more difficult problem is that of finding accurate lower bounds. One of the main purposes of this monograph is to discuss the method of intermediate problems and its ramifications, which give sequences of indefinitely improvable lower bounds. The theory of intermediate problems is not quite simple, but the effort is certainly compensated by the high-precision numerical results which have been obtained by several scientists.

There are of course other methods for the numerical determination of upper and lower bounds, such as inclusion theorems going back to Temple's formula, etc. Some of these methods, while not giving improvable approximations, have the advantage of extreme simplicity as far as the theory is concerned, which may explain their popularity. It would be interesting to compare the range of applicability and the numerical results of the various methods in order to have a clear picture of their efficiencies. There is still a great variety of eigenvalue problems of importance for pure mathematics, mechanics, engineering, and quantum physics which could provide the real challenge and test of any method. While a complete discussion of other methods is beyond the scope of this book, we shall mention some of them as their connections with intermediate problems arise.

While some of the methods discussed in this monograph were originally intended as computational devices, they have led, somewhat unexpectedly, to many theoretical results, in seemingly disconnected fields, which became of great interest by themselves independently of any numerical application.

Chapters 1 and 2 deal with the classical variational theory of eigenvalues and the Rayleigh–Ritz method for upper bounds. In Chapters 4–6, we develop the first and second types of intermediate problems with their variants and numerical applications. Theoretical ramifications of intermediate problems are given in Chapters 3 and 7–9. Other important theoretical results that are also outgrowths of intermediate problems but are not included in the present monograph are mentioned briefly at the end of Chapter 9.

An exposition of the method of intermediate problems has been given in Gould's book [12]. A Russian translation [12a] of the second edition of Gould's book appeared in 1970. Intermediate problems have also been discussed in Fichera's book [6], in the second Russian edition of Mikhlin's book [22] and in some expository articles, including [BF2, D2, FR1, W15, W32]. The present monograph has points in common with these earlier works. However, we bring out new viewpoints and interconnections not previously discussed. Moreover, we give an exposition of several developments (both numerical and theoretical) which have been published in just the last five or six years.

We refer the reader also to Appendix A where some further details are given, discussion of which would be premature in this Introduction.

Chapter One

The Variational Characterization of Eigenvalues

1. A Preliminary Survey of the Classical and Minimax Principles for Eigenvalues

In this section, we outline the development of the variational principles for eigenvalues, which will be discussed and used in subsequent parts of the book. As already mentioned in the Introduction, the history of this development is quite involved and seemingly full of misunderstandings, which we do not claim to resolve completely.

The century-old variational characterization of eigenvalues is due to Weber [W1] and Lord Rayleigh [25]. For higher eigenvalues, this is sometimes called the recursive formulation because in the definition of the higher eigenvalues the eigenvectors corresponding to all previous eigenvalues appear explicitly.

Later, some variational definitions for higher eigenvalues, which do not use other eigenelements in such an obvious way as in the classical case, were introduced. Nearly a quarter of a century after Weber and Rayleigh gave the classical principle, Poincaré obtained a set of inequalities (or bounds) for a finite but arbitrary number of eigenvalues. These results, which he mentioned in a somewhat casual way in his celebrated paper on partial differential equations [P4], were all but forgotten and today are usually, but incorrectly, attributed to Ritz. Poincaré's inequalities were independently rediscovered for finite-dimensional spaces by Fischer [F8] in 1905, who in an ingenious way formulated his result as a *minimum–maximum principle*. It should be noted that Fischer's proof was algebraic rather than variational. For operators in infinite-dimensional spaces, the inequalities of Poincaré were formulated as a *minimum–maximum principle* by Pólya [P5], who clearly recognized Fischer's contribution as well. In many other presentations, this principle is confused with the *maximum–minimum principle*, which will be discussed below. Before leaving this subject, let us remark that Poincaré's bounds for higher eigenvalues (as given for a membrane fixed on the boundary) should not be confused with his other fundamental inequality concerning the second eigenvalue of a free membrane, which also appears in the same paper [P4].

An even more important variational characterization, *the maximum–minimum principle*, is claimed by Weyl, who used some of its consequences in his famous theory of asymptotic distribution of eigenvalues [W31, W32]. Later, Courant applied the principle contained in Weyl's fundamental inequality to a fairly general typical situation [C2].

These two minimax-type principles are essentially different and, as we shall see, are actually complementary to each other. Still another characterization, the *maxi-mini-max principle*, has recently been introduced [S8], which contains all of the older principles as special cases.

2. Operators of Class $\mathscr{S}$

Let $\mathfrak{H}$ be a real or complex Hilbert space having the scalar product (u, v) and the corresponding norm $\|u\| = (u, u)^{1/2}$. Let A be a selfadjoint linear operator on a subspace $\mathfrak{D}$ dense in $\mathfrak{H}$. We shall always assume that A is bounded below and that the lower part of its spectrum consists of a finite or infinite number of isolated eigenvalues $\lambda_1 \leq \lambda_2 \leq \cdots$ each having finite multiplicity. Let λ_∞ denote the lowest point (if any) in the essential spectrum of A (see Definition A.27†). If $\mathfrak{H}$ is finite-dimensional, or if $\mathfrak{H}$ is infinite-dimensional and $\lambda_k \to \infty$, there is no essential spectrum. In all other cases, the point λ_∞ could be a nonisolated eigenvalue of finite or infinite multiplicity, an isolated eigenvalue of infinite multiplicity, or a spectral point which is not an eigenvalue. There may be, of course, point eigenvalues, even isolated point eigenvalues, which are above λ_∞. However, when we speak about the eigenvalues of A, we shall *always* mean the isolated eigenvalues that are below λ_∞. We shall denote by $u_1, u_2, \ldots$ a corresponding orthonormal sequence of eigenvectors and by $R(u)$ the Rayleigh quotient $(Au, u)/(u, u)$, $u \in \mathfrak{D}$, $u \neq 0$.

The selection of such a class of operators, which we denote by $\mathscr{S}$, is motivated by the fact that many problems in classical mechanics, for

† This notation indicates Definition 27 of Appendix A; this form of cross reference will be used throughout this book.

instance vibrations and buckling, lead to compact selfadjoint operators via the use of Green's functions or to selfadjoint differential operators of type $\mathscr{S}$. More recently in quantum theory, it has been shown that the Schrödinger operators for hydrogen, helium, etc., also yield selfadjoint, semibounded operators of the type described above.

The assumption that the eigenvalues are ordered in an ascending sequence is made only for the sake of convenience and uniformity. Obviously, a descending sequence of isolated eigenvalues that are *above* all limit points of the spectrum can be handled by considering the operator $-A$ instead of A and reversing the roles of words such as maximum, minimum, etc. In view of the spectral theorem, we feel consideration of an ascending sequence to be more natural. In some applications, for instance positive-definite integral operators, the eigenvalues are ordered in the opposite sense. Of course, the signs of the eigenvalues do not play a part.

Other notations, symbols, definitions, etc., which are used throughout the book are given in Appendix A.

3. Rayleigh's Principle and the Classical Characterization

The starting point in any discussion of the variational theory of eigenvalues is the following principle, which is the oldest characterization of eigenvalues as minima.

Theorem 1. *The eigenvalues of $A \in \mathscr{S}$ are given by the equations*

$$\lambda_1 = \min_{u \in \mathfrak{D}} R(u) \tag{1}$$

and

$$\lambda_n = \min_{\substack{u \in \mathfrak{D} \\ (u, u_j) = 0 \\ j = 1, 2, \ldots, n-1}} R(u) \qquad (n = 2, 3, \ldots). \tag{2}$$

Equation (1) is usually called *Rayleigh's principle*, and Eq. (2) was first given by Weber [W1].

There are several proofs for compact operators to be found in many textbooks; see, for instance, [3, 26]. Since A is not necessarily compact here, we give a proof which covers all cases. In the following, we use general spectral theory as developed in the late 1920's by J. von Neumann, E. Schmidt, and M. H. Stone. Without using the spectral theorem, we can obtain the weaker result given in Theorem 2.

Proof. In order to show that (1) holds, let E_λ denote the spectral resolution of the identity for A. Since we have assumed that λ_1 is the initial point of $\sigma(A)$, it follows that $E_\lambda = 0$ for $\lambda < \lambda_1$. Therefore, for any $u \in \mathfrak{D}$, the inequality

$$(Au, u) = \int_{-\infty}^{\infty} \lambda \, d(E_\lambda u, u) = \int_{\lambda_1 - 0}^{\infty} \lambda \, d(E_\lambda u, u) \geq \lambda_1(u, u) \tag{3}$$

is satisfied. However, $R(u_1) = \lambda_1$, and so we obtain (1).

Let us note that if A were a compact operator, the integrals in Eq. (3) would be replaced by infinite series (or sums) and we would have

$$\begin{aligned}(Au, u) &= \sum_{j=1}^{\infty} \lambda_j^-(u, u_j^-)u_j^- + \sum_{j=1}^{\infty} \lambda_j^+(u, u_j^+)u_j^+ \\ &\geq \lambda_1^- \left[\sum_{j=1}^{\infty} (u, u_j^-)u_j^- + \sum_{j=1}^{\infty} (u, u_j^+)u_j^+\right] \\ &= \lambda_1^-(u, u),\end{aligned} \tag{4}$$

where $\lambda_1^- \leq \lambda_2^- \leq \cdots$ and $\lambda_1^+ \geq \lambda_2^+ \geq \cdots$ are the negative and positive eigenvalues of A, and $u_1^-, u_2^-, \ldots$ and $u_1^+, u_2^+, \ldots$ are the corresponding eigenvectors.

Similarly, to prove that (2) holds, for fixed $n = 2, 3, \ldots$, we consider $u \in \mathfrak{D}$ such that

$$(u, u_j) = 0 \qquad (j = 1, 2, \ldots, n-1). \tag{5}$$

Of course, we tacitly assume that we have at least n eigenvalues. Then, it follows from the spectral theorem that, for $\lambda < \lambda_n$, $E_\lambda u$ is a linear combination of eigenvectors corresponding to eigenvalues not greater than λ_n. In view of (5), we must have $E_\lambda u = 0$, which yields the inequality

$$(Au, u) = \int_{\lambda_n - 0}^{\infty} \lambda \, d(E_\lambda u, u) \geq \lambda_n(u, u).$$

However, the vector u_n satisfies the orthogonality conditions (5) and the equality $R(u_n) = \lambda_n$, so we have characterization (2).

Let us now give a related theorem, which is proved without the use of the spectral theorem but with the additional assumption that the minima exist.

Theorem 2. *Let $A \in \mathcal{S}$ have the property that there exist vectors $v_1, v_2, \ldots \in \mathfrak{D}$ such that*

$$\mu_1 = R(v_1) = \min_{u \in \mathfrak{D}} R(u)$$

and

$$\mu_n = R(v_n) = \min_{\substack{u \in \mathfrak{D} \\ (u, v_j) = 0 \\ j = 1, 2, \ldots, n-1}} R(u)$$

for $n = 1, 2 \ldots$. Then, every minimum μ_j ($j = 1, 2, \ldots$) is an eigenvalue of A and for every eigenvalue λ_j of A we have $\lambda_j = \mu_j$ ($j = 1, 2, \ldots$). Moreover, each v_j is an eigenvector corresponding to λ_j.

All negative-semidefinite compact operators and all differential operators having positive-definite compact inverses have the above property.

PROOF. We first show that every minimum is an eigenvalue. By construction, the vectors $\{v_i\}$ are orthogonal and without loss of generality we may take them to be normalized. Let ε be any real or complex scalar and let w be any vector in $\mathfrak{D}$. Then, since μ_1 is the minimum, we have

$$(A[v_1 + \varepsilon w], v_1 + \varepsilon w) \geq \mu_1(v_1 + \varepsilon w, v_1 + \varepsilon w).$$

Expanding the inner products and using the fact that $(Av_1, v_1) = \mu_1(v_1, v_1)$, we obtain the inequality

$$0 \leq |\varepsilon|^2 [(Aw, w) - \mu_1(w, w)] + 2 \operatorname{Re}\{\bar{\varepsilon}(Av_1 - \mu_1 v_1, w)\}$$

for any scalar ε. Such an inequality could hold for arbitrary values of ε only if $(Av_1 - \mu_1 v_1, w) = 0$. Since w is any arbitrary vector in the dense subspace $\mathfrak{D}$, it follows that

$$Av_1 = \mu_1 v_1$$

and, therefore, μ_1 is an eigenvalue. Now, let j be any fixed index ($j = 2, 3, \ldots$), assume that $\mu_1, \mu_2, \ldots, \mu_{j-1}$ are eigenvalues, and consider μ_j. Again let ε be any scalar and now let w be any vector in $\mathfrak{D}$ such that

$$(w, v_i) = 0 \qquad (i = 1, 2, \ldots, j - 1). \tag{6}$$

Proceeding as we did above, we have the inequality

$$(A[v_j + \varepsilon w], v_j + \varepsilon w) \geq \mu_j(v_j + \varepsilon w, v_j + \varepsilon w),$$

which leads to the condition

$$(Av_j - \mu_j v_j, w) = 0$$

for all w satisfying (6). Therefore, it follows that

$$Av_j - \mu_j v_j = \alpha_1 v_1 + \alpha_2 v_2 + \cdots + \alpha_{j-1} v_{j-1}$$

for some scalars $\alpha_1, \alpha_2, \ldots, \alpha_{j-1}$. Taking the inner product of both sides with v_i ($i = 1, 2, \ldots, j - 1$), we obtain the equations

$$\alpha_i = (Av_j, v_i) = (v_j, Av_i) = \mu_i(v_j, v_i) = 0,$$

so that

$$Av_j = \mu_j v_j,$$

which means that every minimum is an eigenvalue.

We now show that every eigenvalue is a minimum. In fact, since μ_1 is an eigenvalue, it follows that $\lambda_1 \leq \mu_1$. However

$$\mu_1 = \min_{u \in \mathfrak{D}} R(u) \leq R(u_1) = \lambda_1,$$

so that $\lambda_1 = \mu_1$. Suppose that $\lambda_1 = \mu_1$, $\lambda_2 = \mu_2, \ldots, \lambda_{j-1} = \mu_{j-1}$. Since μ_j is an eigenvalue, we have $\lambda_j \leq \mu_j$. On the other hand,

$$\mu_j = \min_{\substack{u \in \mathfrak{D} \\ (u, v_i) = 0 \\ i = 1, 2, \ldots, j-1}} R(u) = \min_{\substack{u \in \mathfrak{D} \\ (u, u_i) = 0 \\ i = 1, 2, \ldots, j-1}} R(u) \leq R(u_j) = \lambda_j,$$

which proves that $\lambda_j = \mu_j$.

REMARK. It should be noted that usually Theorem 2 is proved under the additional assumption $\lambda_k \to \infty$; see [3, p. 494].

Chapter Two

The Rayleigh–Ritz Method

1. Poincaré's Inequalities: The Theoretical Foundation of the Rayleigh–Ritz Method

We now develop the inequalities of Poincaré, which in our terminology relate the eigenvalues of an operator of class $\mathscr{S}$ to the eigenvalues of its projection on a finite-dimensional subspace (see Definition A.19). Incidentally, this presentation will provide a simplification of the older proofs.

Let $\mathfrak{V}_n$ (or, for short, $\mathfrak{V}$) be any n-dimensional supspace of $\mathfrak{D}$ and let V_n (or V) be the orthogonal projection operator onto $\mathfrak{V}$. Consider the equation

$$VAVu = \Lambda u. \tag{1}$$

By construction, VAV is hermitian (or symmetric) and has all of $\mathfrak{H}$ for its domain. Therefore, by Definitions A.10 and A.11, VAV is selfadjoint. Moreover, VAV is also a transformation from the finite-dimensional space $\mathfrak{V}$ into itself, so that the spectrum of VAV, considered as an operator on $\mathfrak{V}$ consists of n real eigenvalues, $\Lambda_1 \leq \Lambda_2 \leq \cdots \leq \Lambda_n$, some of which may be zero. Denote by $w_1, w_2, \ldots, w_n$ the corresponding orthonormal set of eigenvectors in $\mathfrak{V}$. The eigenvalues $\{\Lambda_i\}$ and eigenvectors $\{w_i\}$ will be called *nontrivial.* If, however, we consider VAV as an operator on $\mathfrak{H}$, its spectrum consists of the eigenvalues $\Lambda_1, \Lambda_2, \ldots, \Lambda_n$ as well as the eigenvalue $\lambda = 0$. If $\mathfrak{H}$ is infinite-dimensional, zero is an eigenvalue of infinite multiplicity.

Theorem 1. (*Poincaré*) *For any n-dimensional space $\mathfrak{V}_n$, the eigenvalues $\{\Lambda_i\}$ of VAV satisfy the inequalities*

$$\lambda_1 \leq \Lambda_1, \quad \lambda_2 \leq \Lambda_2, \ldots, \quad \lambda_n \leq \Lambda_n. \tag{2}$$

PROOF. For a given index j $(1 \leq j \leq n)$, we consider a *test function*

$$w_0 = \alpha_1 w_1 + \alpha_2 w_2 + \cdots + \alpha_j w_j \neq 0,$$

where $\alpha_1, \alpha_2, \ldots, \alpha_j$ are adjusted so that the $j-1$ orthogonality conditions $(w_0, u_i) = 0$ $(i = 1, 2, \ldots, j-1)$ are satisfied. Without loss of generality, we can assume that $\|w_0\| = 1$.

It follows from Eq. (1.3.2)† that

$$\lambda_j = \min_{\substack{u \in \mathfrak{D} \\ (u, u_i) = 0 \\ (i = 1, 2, \ldots, j-1)}} R(u) \leq R(w_0).$$

On the other hand, since $Vw_0 = w_0$ and $(w_0, w_0) = 1$, we have

$$R(w_0) = \frac{(Aw_0, w_0)}{(w_0, w_0)} = (VAVw_0, w_0) = \sum_{i=1}^{j} |\alpha_i|^2 \Lambda_i \leq \Lambda_j.$$

The original proof of Theorem 1 by Poincaré [P4], repeated by Pólya and Schiffer [PS1], is somewhat more complicated than ours due to the fact that Poincaré began with an arbitrary basis $v_1, v_2, \ldots, v_n$ for $\mathfrak{B}_n$.

The inequalities of Poincaré are of theoretical importance in that they lead to an independent characterization of eigenvalues, *the minimum–maximum principle.*

These inequalities (2) also have an important practical application, usually called the *Rayleigh–Ritz method,* which we shall discuss in detail in Section 3.

2. The Minimum–Maximum Principle

While the classical characterization is very important as a theoretical tool, it has the disadvantage that it cannot be used to determine higher eigenvalues without using explicitly all preceding eigenvectors. The following approach overcomes this difficulty.

Let us focus our attention on the inequality

$$\lambda_n \leq \Lambda_n \tag{1}$$

from the set of inequalities (1.2). Since each $v \in \mathfrak{B}_n$ can be written as a linear combination

$$v = \beta_1 w_1 + \beta_2 w_2 + \cdots + \beta_n w_n$$

† That is, Eq. (2) of Chapter 1, Section 3. This form of notation is used throughout the book.

it follows that

$$(Av, v) = \sum_{i=1}^{n} |\beta_i|^2 \Lambda_i \leq \Lambda_n(v, v) \tag{2}$$

for any choice of v. Moreover, the particular choice $v = w_n$ yields the equality $R(v) = \Lambda_n$. Combining this equality with inequality (1) we see that

$$\lambda_n \leq \Lambda_n = \max_{u \in \mathfrak{V}_n} R(u).$$

We can now give a characterization of λ_n which is independent in the sense that it does not explicitly use the eigenvectors to characterize the eigenvalues.

Theorem 1. *(Fischer–Pólya) The minimum–maximum principle. Let $\mathfrak{V}_n$ denote any n-dimensional subspace of $\mathfrak{D}$ $(n = 1, 2, \ldots)$. Then, the eigenvalues of A are characterized by the equation*

$$\lambda_n = \min_{\mathfrak{V}_n} \max_{u \in \mathfrak{V}_n} R(u). \tag{3}$$

PROOF. From the above discussion, we have

$$\lambda_n \leq \max_{u \in \mathfrak{V}_n} R(u) \tag{4}$$

for any $\mathfrak{V}_n \subset \mathfrak{D}$, $\dim \mathfrak{V}_n = n$. Let us now consider the subspace $\mathfrak{U}_n = \mathrm{sp}\,\{u_1, u_2, \ldots, u_n\}$, where sp { } denotes the subspace consisting of all linear combinations of the vectors enclosed. In this case, we have $R(v) \leq \lambda_n$, for any $v \in \mathfrak{U}_n$, and $R(u_n) = \lambda_n$. Therefore,

$$\lambda_n = \max_{u \in \mathfrak{U}_n} R(u). \tag{5}$$

Using inequality (4) and equality (5), we obtain the characterization (3).

We can take into account the possibility that the multiplicity of a given λ_n is greater than 1 in the following more explicit way.

Theorem 2. *For a fixed n, let m(n) (for short, m) be defined by*

$$m = \min\{j \,|\, \lambda_j = \lambda_n\} \tag{6}$$

and let $\mathcal{G}_n$ be the set of all subspaces $\mathfrak{V} \subset \mathfrak{D}$ such that $m \leq \dim \mathfrak{V} < \infty$. Then, the eigenvalues of A are given by the equation

$$\lambda_n = \min_{\mathfrak{V} \in \mathcal{G}_n} \max_{u \in \mathfrak{V}} R(u). \tag{7}$$

PROOF. For any $\mathfrak{V} \in \mathscr{G}_n$, let $r = \dim \mathfrak{V}$. Since $r \geq m$, we have the inequality

$$\lambda_n = \lambda_m \leq \Lambda_m \leq \Lambda_r \leq \max_{u \in \mathfrak{V}} R(u).$$

It is clear that $\mathfrak{U}_n \in \mathscr{G}_n$, and thus the equality

$$\lambda_n = \max_{u \in \mathfrak{U}_n} R(u)$$

is obtained, which proves (7).

3. Upper Bounds for Eigenvalues

Based upon the theoretical investigation just discussed in Sections 1 and 2, a famous method for obtaining upper bonds for eigenvalues has been developed, which is known today as the *Rayleigh–Ritz method.* This method provides a straightforward and efficient means of computing non-increasing upper bounds for an arbitrary but finite number of eigenvalues of any operator in class $\mathscr{S}$.

The main idea of the method is to restrict a given operator to a finite-dimensional subspace of its domain, yielding a matrix problem for which the eigenvalues are numerically computable. It then follows from Poincaré's result (1.2) that the computed eigenvalues are upper bounds for the initial eigenvalues of the given operator.

Let us now describe the method in detail. Consider an operator $A \in \mathscr{S}$ for which the eigenvalues $\lambda_1 \leq \lambda_2 \leq \cdots$ are unknown. For an arbitrary integer n, choose any n linearly independent vectors $v_1, v_2, \ldots, v_n \in \mathfrak{D}$. In applications, these vectors are usually called *test functions* and satisfy the boundary conditions of the original problem. Such vectors generate a subspace $\mathfrak{V}_n$. If we now restrict the operator A to an operator from $\mathfrak{V}_n$ to $\mathfrak{V}_n$, we have the eigenvalue problem

$$V_n A V_n u = \Lambda u, \tag{1}$$

where V_n is the orthogonal projection onto $\mathfrak{V}_n$. From (1.2), the eigenvalues $\Lambda_1 \leq \Lambda_2 \leq \cdots \leq \Lambda_n$ satisfy the inequalities

$$\lambda_1 \leq \Lambda_1, \quad \lambda_2 \leq \Lambda_2, \ldots, \quad \lambda_n \leq \Lambda_n.$$

It remains to solve explicitly for $\Lambda_1, \Lambda_2, \ldots, \Lambda_n$. For the moment, let us introduce an orthonormal basis v_1' v_2', $\ldots$, v_n' for $\mathfrak{V}_n$. For $u \in \mathfrak{V}_n$, we have $u = V_n u$, so that we can write

$$u = V_n u = \sum_{i=1}^{n} (u, v_i')v_i',$$
$$AV_n u = \sum_{i=1}^{n} (u, v_i')Av_i', \tag{2}$$

and

$$V_n AV_n u = \sum_{j=1}^{n} \sum_{i=1}^{n} (u, v_i')(Av_i', v_j')v_j'. \tag{3}$$

Substituting (2) and (3) in (1), we obtain the equation

$$\sum_{j=1}^{n} \sum_{i=1}^{n} (u, v_i')(Av_i', v_j')v_j' - \Lambda \sum_{i=1}^{n} (u, v_i')v_i' = 0. \tag{4}$$

Taking inner products of both sides of (4) with v_k', we have the set of equations

$$\sum_{i=1}^{n} (u, v_i')(Av_i', v_k') - \Lambda(u, v_k') = 0 \qquad (k = 1, 2, \ldots, n). \tag{5}$$

The system (5) is a matrix eigenvalue problem which has nontrivial solutions if and only if

$$\det\{(Av_i', v_k') - \Lambda\, \delta_{ik}\} = 0 \qquad (i, k = 1, 2, \ldots, n). \tag{6}$$

We now transform Eq. (6) into an equation that involves only the arbitrary basis $v_1, v_2, \ldots, v_n$. In fact, we have

$$v_i' = \sum_{j=1}^{n} \beta_{ij} v_j \qquad (i = 1, 2, \ldots, n), \tag{7}$$

where $\{\beta_{ij}\}$ is a nonsingular matrix. Therefore, we can write

$$(Av_i', v_k') = \sum_{j=1}^{n} \beta_{ij} \sum_{h=1}^{n} \beta_{kh}(Av_j, v_h) \tag{8}$$

and

$$\delta_{ik} = (v_i', v_k') = \sum_{j=1}^{n} \beta_{ij} \sum_{h=1}^{n} \beta_{kh}(v_j, v_h). \tag{9}$$

Using Eqs. (8) and (9) in (6), we see that

$$\det\{(Av_i', v_k') - \Lambda\, \delta_{ik}\} = \det[\{\beta_{ij}\}\{(Av_j, v_h) - \Lambda(v_j, v_h)\}\{\beta_{hk}^T\}].$$

Since $\{\beta_{ij}\}$ is nonsingular, the system (5) has a nontrivial solution if and only if

$$\det\{(Av_j, v_h) - \Lambda(v_j, v_h)\} = 0 \qquad (j, h = 1, 2, \ldots, n). \tag{10}$$

Equation (10) in principle gives an explicit determination of the eigenvalues $\Lambda_1, \Lambda_2, \ldots, \Lambda_n$.

A natural question which arises in the consideration of this method is whether or not the upper bounds $\Lambda_1, \Lambda_2, \ldots, \Lambda_n$ are improved (that is, decreased) by taking more vectors, say $v_{n+1}, v_{n+2}, \ldots, v_{n+r}$. The answer is that the new eigenvalues are not "worse" upper bounds than the old eigenvalues. To be more precise, let $\Lambda_1 \le \Lambda_2 \le \cdots \le \Lambda_n$ and $w_1, w_2, \ldots, w_n$ denote the eigenvalues and eigenvectors of (1) and let $\Lambda_1' \le \Lambda_2' \le \cdots \le \Lambda_n'$ denote the first n eigenvalues of $V_{n+r}AV_{n+r}u = \Lambda u$. Then, from the minimum–maximum principle, the jth eigenvalue ($j = 1, 2, \ldots, n$) of $V_{n+r}AV_{n+r}$ is given by

$$\Lambda_j' = \min_{\mathfrak{V}_j \subset \mathfrak{V}_{n+r}} \max_{u \in \mathfrak{V}_j} R(u).$$

For the particular subspace $\mathfrak{W}_j = sp\{w_1, w_2, \ldots, w_j\}$, we have

$$\Lambda_j' \le \max_{u \in \mathfrak{w}_j} R(u) = \Lambda_j,$$

which means that the upper bounds are not increased by increasing the dimension of the finite-dimensional subspace. However, there is no guarantee that the bounds are better. There is no improvement, for instance, if we take eigenfunctions as test functions.

REMARK. There is no doubt that the inequality

$$\lambda_1 \le R(u) \qquad (u \in \mathfrak{D})$$

was given by Rayleigh. Although the inequalities for higher eigenvalues are usually attributed to Ritz, it has been noted that seemingly both Rayleigh and Ritz were familiar only with the inequality $\lambda_1 \le \Lambda_1$ [W10, P5]. In fact, in the paper of Ritz [R2] where he is supposed to have obtained upper bounds for 17 eigenvalues for a free plate, he actually has only found 17 upper bounds for λ_1. In order to have upper bounds for 17 eigenvalues, Ritz would have been required to consider the determinant of a 17×17 matrix, but we were unable to find in Ritz's papers such a determinant, which could not be easily overlooked. However, since Ritz emphasized the numerical applications, although without sufficient theoretical foundation, there is a historical justification for calling this result the *Rayleigh–Ritz method.*

Let us mention that the equations of the Rayleigh–Ritz method can also be obtained in the following way, which is familiar in applications. Consider the minimum of (Av, v) for all $v = \alpha_1 v_1 + \alpha_2 v_2 + \cdots + \alpha_n v_n$ for fixed test functions $v_1, v_2, \ldots, v_n \in \mathfrak{D}$ under the side condition $(v, v) = 1$.

(The choice of the test functions is left to the discretion of the computer.) Introduce a Lagrange multiplier Λ and write the equations

$$(\partial/\partial\alpha_k)[(Av, v) - \Lambda(v, v)] = 0 \qquad (k = 1, 2, \ldots, n). \tag{11}$$

Then, the determinant of (11) is exactly (10). Of course, this procedure would not of itself yield the results given in this section for higher eigenvalues.

4. A Necessary and Sufficient Criterion in the Minimum–Maximum Theory

As in the case of the new maximum–minimum theory [W18], which was actually developed earlier, the classical choice of the subspace, by which we mean here $\mathfrak{V} = \mathfrak{U}_n$, is not the only case for which the minimum is attained in (2.4). We shall now give a set of necessary and sufficient conditions on the subspace $\mathfrak{V}$ for equality (2.5) to hold [S5]. They are much simpler than the corresponding conditions for the maximum–minimum theory (see Section 7.4) which were given earlier.

Theorem 1. *A necessary condition for the equality*

$$\lambda_n = \max_{u \in \mathfrak{V}} R(u) \tag{1}$$

to hold is that the dimension r of $\mathfrak{V}$ satisfy the inequality

$$r \leq M(n), \tag{2}$$

where $M(n)$ (or, for short, M) is defined by

$$M = \max\{j \mid \lambda_j = \lambda_n\}. \tag{3}$$

Let $v_1, v_2, \ldots, v_r$ denote any basis for $\mathfrak{V}$. Assuming that (2) *holds, we have the following necessary and sufficient condition for equality* (1)*: the quadratic form with the symmetric matrix*

$$\{([A - \lambda_n I]v_i, v_k)\} \qquad (i, k = 1, 2, \ldots, r) \tag{4}$$

is negative-semidefinite.

Proof. In order to establish our necessary condition, suppose that equality (1) holds for some $\mathfrak{V}$ such that

$$M < r = \dim \mathfrak{V}. \tag{5}$$

Applying Theorem 2.1, we have

$$\lambda_r \leq \max_{u \in \mathfrak{V}} R(u) = \lambda_n = \lambda_M, \tag{6}$$

which contradicts inequality (5).† In developing the proof of the necessary and sufficient condition, let us for the moment consider as our basis for $\mathfrak{V}$ the set of eigenvectors $w_1, w_2, \ldots, w_r$ of the operator VAV. In this case, matrix (4) becomes the diagonal matrix

$$\begin{Bmatrix} \Lambda_1 - \lambda_n & 0 & 0 & \cdots & 0 \\ 0 & \Lambda_2 - \lambda_n & 0 & \cdots & 0 \\ \vdots & & & & \\ 0 & 0 & & & \Lambda_r - \lambda_n \end{Bmatrix}. \tag{7}$$

If equality (1) holds, it follows that

$$\Lambda_i = (Aw_i, w_i) = R(w_i) \le \lambda_n \qquad (i = 1, 2, \ldots, r). \tag{8}$$

Therefore, the quadratic form defined by matrix (7) is negative-semi-definite. Conversely, if the quadratic form of matrix (7) is negative-semi-definite, then we have (8). Since every $u \in \mathfrak{V}$, $(u \ne 0)$ can be written as

$$u = \sum_{i=1}^{r} \gamma_i w_i,$$

we have

$$R(u) = \left\{\sum_{i=1}^{r} |\gamma_i|^2 (Aw_i, w_i)\right\} \Big/ \left\{\sum_{i=1}^{r} |\gamma_i|^2\right\} \le \lambda_n. \tag{9}$$

Combining inequalities (2.4) and (9), we obtain

$$\lambda_n \le \max_{u \in \mathfrak{V}} R(u) \le \lambda_n,$$

so that equality (1) holds. Let us now obtain the criterion for a general basis $v_1, v_2, \ldots, v_r$. Since $\{v_i\}$ and $\{w_i\}$ are both bases for the space $\mathfrak{V}$, there exists a matrix $\{\alpha_{ij}\}$ such that

$$v_i = \sum_{j=1}^{r} \alpha_{ij} w_j \qquad (i = 1, 2, \ldots, r).$$

Therefore, matrix (4) is equal to

$$\left\{\left([A - \lambda_n I] \sum_{j=1}^{r} \alpha_{ij} w_j, \sum_{h=1}^{r} \alpha_{kh} w_h\right)\right\}$$
$$= \{\alpha_{ij}\}\{([A - \lambda_n I]w_j, w_h)\}\{\alpha_{hk}^T\} \qquad (i, k = 1, 2, \ldots, r).$$

† It could happen that λ_M is the largest isolated eigenvalue at the beginning of the spectrum, in which case λ_∞ would play the role of λ_r in (6); see Theorem 3.

From this, we see that a change in basis of $\mathfrak{V}$ has the same effect as changing the basis upon which matrix (7) is represented. Therefore, the quadratic form given by matrix (4) will be negative-semidefinite if and only if the quadratic form given by (7) is negative-semidefinite.

REMARK. It should be pointed out that in the above criterion the value λ_n appears explicitly. This is not surprising, since even in the classical theory (1.3.2), which gives only sufficient conditions for equality, the eigenvectors $u_1, u_2, \ldots, u_n$ appear explicitly (see also Chapter 7).

In some of the earlier papers on this subject [S5, S8], the criterion was formulated in the following way.

Theorem 2. *A necessary condition for* (1) *to hold is that the dimension r of* $\mathfrak{V}$ *satisfy* (2). *Assuming that* (2) *holds, a necessary and sufficient condition for* (1) *to hold is that for every* $\varepsilon \geqq 0$ *the quadratic form*

$$\{([A - (\lambda_n + \varepsilon)I]v_i, v_k)\} \qquad (i, k = 1, 2, \ldots, r)$$

be negative-semidefinite.

The proof is virtually identical to the proof given above. The criterion with $\varepsilon \geq 0$ most closely resembles the analogous criterion for the earlier new maximum–minimum theory, as we shall see in Section 7.4.

EXAMPLE 1. Using our criterion, we can now give a nonclassical subspace $\mathfrak{V}$ for which equality (1) holds. Let $\lambda_1 < \lambda_2 < \lambda_3$ and $m = n = M = 2$. Consider the subspace $\mathfrak{V} = sp\{u_2, u_1 + \beta u_3\}$, where

$$0 < |\beta|^2 \leq (\lambda_2 - \lambda_1)/(\lambda_3 - \lambda_2). \tag{10}$$

Every $v \in \mathfrak{V}$ $[(v, v) = 1]$ can be written as

$$v = \xi u_1 + \eta u_2 + \xi\beta u_3,$$

where $|\xi|^2 + |\eta|^2 + |\xi|^2|\beta|^2 = 1$. Therefore, we have, by (10),

$$\begin{aligned} R(v) &= |\xi|^2\lambda_1 + |\eta|^2\lambda_2 + |\xi|^2|\beta|^2\lambda_3 \\ &= |\xi|^2\lambda_1 + (1 - |\xi|^2 - |\xi|^2|\beta|^2)\lambda_2 + |\xi|^2|\beta|^2\lambda_3 \\ &= |\xi|^2(\lambda_1 - \lambda_2) + \lambda_2 + |\xi|^2|\beta|^2(\lambda_3 - \lambda_2) \\ &\leq |\xi|^2(\lambda_1 - \lambda_2) + \lambda_2 + |\xi|^2(\lambda_2 - \lambda_1) = \lambda_2 . \end{aligned} \tag{11}$$

This implies

$$\max_{v \in \mathfrak{V}} R(v) \leq \lambda_2 .$$

However, from (2.4), we have

$$\lambda_2 \leq \max_{u \in \mathfrak{V}} R(v)$$

and therefore

$$\lambda_2 = \max_{v \in \mathfrak{V}} R(v) \tag{12}$$

holds. Observe that when $\beta = 0$ this example reduces to the classical case $\mathfrak{V} = \mathfrak{U}_2$, in which case (12) is trivially satisfied. When $|\beta|^2 = (\lambda_2 - \lambda_1)/(\lambda_3 - \lambda_2)$, we have equality throughout (11), and therefore

$$\lambda_2 = R(v)$$

for all $v \in \mathfrak{V}$, $v \neq 0$. This is the limiting case, since we have

$$R(u_1 + \beta u_3) > \lambda_2 \qquad \text{if} \qquad |\beta|^2 > (\lambda_2 - \lambda_1)/(\lambda_3 - \lambda_2).$$

For this example, the matrix (7) of Theorem 1 is

$$\begin{Bmatrix} 0 & 0 \\ 0 & (\lambda_1 - \lambda_2) + |\beta|^2(\lambda_3 - \lambda_2) \end{Bmatrix},$$

which is clearly negative-semidefinite for all β satisfying the inequalities (10).

One might assume that an example such as the above is somewhat exceptional. Surprisingly, it turns out, as in Section 7.6, that there exists a "nonclassical" choice in all cases, with two very special exceptions, which we shall now prove. We begin by giving a more precise definition of a nonclassical choice.

Definition 1. For a given index n ($n = 1, 2, \ldots$), a subspace $\mathfrak{V}_n$ is said to be a nonclassical choice for the minimum of the maximum in (2.3) if $\mathfrak{V}_n \not\subset sp\{u_1, u_2, \ldots, u_{M(n)}\}$.

Then we have the following general result [S13].

Theorem 3. *The only two cases in which there does not exist a nonclassical choice for the minimum of the maximum in* (2.3) *are* (i) *A is an operator on a finite-dimensional space and λ_n is its greatest eigenvalue, and* (ii) $\lambda_n = \lambda_1$, *even if $\mathfrak{H}$ is infinite-dimensional.*

PROOF. If, for a fixed λ_n, there is a point in the spectrum of A, not necessarily an eigenvalue, which is greater than λ_n, then, since λ_n is isolated and the spectrum is closed, we choose the smallest such element, say ξ.

Let $\mathfrak{U} = \text{sp}\{u_1, u_2, \ldots, u_{M(n)}\}$. In view of the existence of ξ, we know that $\mathfrak{U} \neq \mathfrak{D}$. Since $\dim \mathfrak{U} < \infty$ and $\mathfrak{D}$ is dense, it follows from a lemma of Gohberg–Krein (Lemma A.1) that there exists a vector $w \in \mathfrak{U}^\perp \cap \mathfrak{D}$ $[(w, w) = 1]$. Letting $\eta = R(w)$, we note that $\lambda_n < \xi \leq \eta$. If $\lambda_1 < \lambda_n$, we choose β so that

$$0 < |\beta|^2 \leq (\lambda_n - \lambda_1)/(\eta - \lambda_n)$$

and set

$$v_1 = (1 + |\beta|^2)^{-1/2}(u_1 + \beta w)$$

and

$$v_2 = u_2, \quad v_3 = u_3, \ldots, \quad v_n = u_n.$$

Then, we have

$$([A - \lambda_n I]v_1, v_1) = (1 + |\beta|^2)^{-1}[(\lambda_1 - \lambda_n) + |\beta|^2(\eta - \lambda_n)] \leq 0$$

and

$$([A - \lambda_n I]v_j, v_j) = \lambda_j - \lambda_n \leq 0 \qquad (j = 2, 3, \ldots, n).$$

The diagonal matrix $\{([A - \lambda_n I]v_i, v_j)\}$ is negative-semidefinite and, therefore, it follows from Theorem 2 that the minimum in (2.3) is attained for the nonclassical choice $\mathfrak{V}_n = \text{sp}\{v_1, v_2, \ldots, v_n\}$. If λ_n is the greatest eigenvalue of an operator on a finite-dimensional space $\mathfrak{H}_N$, then $\mathfrak{H}_N = \text{sp}\{u_1, u_2, \ldots, u_{M(n)}\}$, so that nonclassical choices do not exist. On the other hand, if $A \in \mathcal{S}$ and $\lambda_1 = \lambda_n$, we assume that $\mathfrak{V}_n$ is a nonclassical choice for (2.3). Then, we have

$$\lambda_1 = \min_{u \in \mathfrak{D}} R(u) \leq \min_{u \in \mathfrak{V}_n} R(u) \leq \max_{u \in \mathfrak{V}_n} R(u) = \lambda_n$$

so that $\mathfrak{V}_n \subset \mathfrak{E}_n$ (where $\mathfrak{E}_n$ denotes the eigenspace of λ_n), which contradicts our assumption that $\mathfrak{V}_n$ is nonclassical and completes the proof.

The conclusion in the case $\lambda_1 = \lambda_n$ above is actually a special case of the following theorem, which will be used in Section 8.6.

Theorem 4. *A necessary and sufficient condition on the space $\mathfrak{V}_r$ of* Theorem 2.1 *that the simultaneous equalities*

$$\lambda_1 = \Lambda_1, \quad \lambda_2 = \Lambda_2, \ldots, \quad \lambda_r = \Lambda_r \tag{13}$$

hold is that $\mathfrak{V}_r$ be generated by eigenvectors corresponding to $\lambda_1, \lambda_2, \ldots, \lambda_r$.

PROOF. The sufficiency of the condition is obvious. To prove necessity, assume that the equalities (13) hold. Then, we have

$$VAVw_i = \lambda_i w_i \qquad (i = 1, 2, \ldots, r).$$

Since $VAVw_1 = \lambda_1 w_1$, it follows that

$$\lambda_1 = \min_{u \in \mathfrak{D}} R(u) = R(w_1).$$

From the variational theory (see Section 1.3), this means that w_1 is an eigenvector corresponding to λ_1. Now, suppose that $w_1, w_2, \ldots, w_k$ are eigenvectors corresponding to $\lambda_1, \lambda_2, \ldots, \lambda_k$ $(1 \leq k < r)$. Then, we have

$$\lambda_{k+1} = \min_{\substack{u \in \mathfrak{D} \\ u \perp w_1, w_2, \ldots, w_k}} R(u) = R(w_{k+1})$$

and therefore by the same reasoning as in the case of w_1, w_{k+1} is an eigenvector corresponding to λ_{k+1}, which completes the proof.

5. The Principle of Monotonicity

We shall now derive an important application of the *minimum–maximum principle*, namely to carry over to all λ_n statements that are evident for the first eigenvalue λ_1. Such an application was first conceived by Weyl [W31], who proved the same results as given below, not from Poincaré's inequalities, but from an important inequality which he introduced for this purpose. Weyl's inequality, which will be discussed in Chapter 3, became the foundation of the classical maximum–minimum theory. His approach has been used nearly everywhere in texts since 1910. The presentation given here seems to be of more recent times and has appeared in [W5, FR1, S11].

Let A' be an operator of class $\mathscr{S}$ such that

$$\mathfrak{D}(A') \subset \mathfrak{D}(A) \tag{1}$$

and

$$(Au, u) \leq (A'u, u) \qquad \text{for all} \quad u \in \mathfrak{D}(A'). \tag{2}$$

Then, we say that A is *dominated by* A' and write $A \leq A'$. It is obvious from Rayleigh's principle (1.3.1) that $\lambda_1 \leq \lambda_1'$. The following theorem shows that every eigenvalue of A is not greater than the corresponding eigenvalue of A'.

Theorem 1. (*Monotonicity principle*) *If A' and A are operators of class $\mathscr{S}$ satisfying conditions* (1) *and* (2), *then the eigenvalues λ_i' and λ_i of A' and A, respectively, satisfy the inequalities*

$$\lambda_i \leq \lambda_i' \qquad (i = 1, 2, \ldots). \tag{3}$$

PROOF. For any index i, consider $\mathfrak{U}_i' = \mathrm{sp}\{u_1', u_2', \ldots, u_i'\}$. Then, by the minimum–maximum principle (2.3), we have

$$\lambda_i \le \max_{u \in \mathfrak{U}_i'}[(Au, u)/(u, u)] \le \max_{u \in \mathfrak{U}_i'}(A'u, u)/(u, u)] = \lambda_i', \tag{4}$$

which yields (3).

In the case $(Au, u) < (A'u, u)$ for all $u \in \mathfrak{D}(A')$, it is clear that in (4) we have a strict inequality, and therefore

$$\lambda_i < \lambda_i' \qquad (i = 1, 2, \ldots). \tag{5}$$

Chapter Three

The Classical Maximum-Minimum Theory and Its Extension to Unbounded Operators

1. Weyl's First Fundamental Lemma

The inequality we shall discuss here is the foundation of the *maximum–minimum principle* in its original form. In his Gibbs Lecture, Weyl calls this result his fundamental lemma [W32]. We prefer to say "Weyl's first fundamental lemma," as another result of Weyl is generally known as Weyl's lemma (see Section 4.14). Let us note that Weyl originally considered only a differential operator connected with the vibrating membrane and not the general problem in Hilbert space.

Let $p_1, p_2, \ldots, p_{n-1}$ be $n-1$ arbitrary (not necessarily independent) vectors in $\mathfrak{H}$. Then, $\mathfrak{P} = \mathrm{sp}\{p_1, p_2, \ldots, p_{n-1}\}$ is a subspace whose dimension is less than or equal to $n-1$. Consider the variational problem

$$\lambda(p_1, p_2, \ldots, p_{n-1}) = \min_{\substack{u \in \mathfrak{D} \\ (u, p_i) = 0 \\ i = 1, 2, \ldots, n-1}} R(u). \tag{1}$$

If the operator A is compact, it is well known that the minimum $\lambda(p_1, p_2, \ldots, p_{n-1})$ exists. In fact, in this case, the only difference from Eq. (1.3.1) is that here we are taking the minumum over a smaller space, namely the closed subspace of vectors orthogonal to $p_1, p_2, \ldots, p_{n-1}$, which is itself a Hilbert space. On the other hand, if A is not compact, the existence of the minimum is not obvious and will be proved in Section 3. For the moment, let us assume that the minimum does exist for the type of operator with which we are dealing. Then, we have the following inequality.

Weyl's Lemma. *For any choice of vectors* $p_1, p_2, \ldots, p_{n-1} \in \mathfrak{H}$, *we have the inequality*

$$\lambda(p_1, p_2, \ldots, p_{n-1}) \leq \lambda_n. \tag{2}$$

PROOF. Following Weyl [W31], we consider a *test function*

$$u_0 = \alpha_1 u_1 + \alpha_2 u_2 + \cdots + \alpha_n u_n \neq 0$$

having n arbitrary parameters $\alpha_1, \alpha_2, \ldots, \alpha_n$ which are adjusted in such a way that the $n-1$ orthogonality conditions $(u_0, p_i) = 0$ $(i = 1, 2, \ldots, n-1)$ are satisfied. Moreover, without loss of generality, we can assume that $\|u_0\| = 1$. Then, we have

$$(Au_0, u_0) = |\alpha_1|^2 \lambda_1 + |\alpha_2|^2 \lambda_2 + \cdots + |\alpha_n|^2 \lambda_n \leq \lambda_n.$$

Therefore, it follows that

$$\lambda(p_1, p_2, \ldots, p_{n-1}) \leq (Au_0, u_0) \leq \lambda_n.$$

Weyl's lemma naturally includes the choice $p_1 = u_1, p_2 = u_2, \ldots, p_{n-1} = u_{n-1}$, which yields the classical equality (1.3.2).

The principle of monotonicity (2.5.3), as mentioned earlier, was first derived by Weyl from the inequality (2). In fact, using the notations of Section 2.5, we have

$$\lambda_i = \min_{\substack{u \in \mathfrak{D}(A) \\ (u, u_j) = 0 \\ j = 1, 2, \ldots, i-1}} \frac{(Au, u)}{(u, u)} \leq \min_{\substack{u \in \mathfrak{D}(A') \\ (u, u_j) = 0 \\ j = 1, 2, \ldots, i-1}} \frac{(A'u, u)}{(u, u)} \leq \lambda_i' \qquad (i = 1, 2, \ldots).$$

2. The Maximum–Minimum Principle

As we have seen, Fischer [F8] introduced the suggestive term *minimum–maximum* in the theory of upper bounds. Paralleling Fischer, we have the following maximum–minimum principle contained in Weyl's inequality (1.2).

The Maximum–Minimum Principle. *The higher eigenvalues of A are characterized by the equation*

$$\lambda_n = \max_{p_1, p_2, \ldots, p_{n-1} \in \mathfrak{H}} \min_{\substack{u \in \mathfrak{D} \\ (u, p_i) = 0 \\ i = 1, 2, \ldots, n-1}} R(u) \tag{1}$$

for $n = 2, 3, \ldots$.

In fact, by (1.2), the minimum of $R(u)$ is less than or equal to λ_n, and by (1.3.2), the maximum is attained for the choice $p_1 = u_1$, $p_2 = u_2, \ldots,$ $p_{n-1} = u_{n-1}$.

The case $n = 1$ may be included in (1) by considering the minimum problem with no orthogonality conditions, in which case we just have the classical result (1.3.1).

In the extensive literature, it is not usually emphasized that the choice $p_i = u_i$ $(i = 1, 2, \ldots, n-1)$ is *only a sufficient condition* for equality in (1.2) and the question of a criterion for equality is not even formulated.

In Chapter 7, we shall discuss *Weinstein's new maximum–minimum theory*, which not only yields Weyl's lemma in a new way, but also gives the necessary and sufficient conditions for equality. While the detailed discussion will be given later, it is of interest to give here some indications of one of the basic ideas of the new theory, because similar ideas will be used in Section 3.

The fundamental starting point (see p. 26) is that the minimum $\lambda(p_1, p_2, \ldots, p_{n-1})$ is the first eigenvalue of a new problem

$$Au - PAu = \lambda u, \qquad Pu = 0, \tag{2}$$

where P is the orthogonal projection operator onto the subspace $\mathfrak{P} =$ sp $\{p_1, p_2, \ldots, p_{n-1}\}$. A detailed study of the spectrum of (2) is given in Chapters 4 and 9.

We shall again consider on several occasions operators of the type appearing in (2). The operator $A - PA$ on the subspace $\mathfrak{Q} = \mathfrak{P}^{\perp}$ called the *part of A in* $\mathfrak{Q}$, is selfadjoint (see Theorem A.3). Here we do not assume that $\mathfrak{Q}$ is an invariant subspace for A; compare [19, pp. 22, 172]. The determination of the spectrum of the part in A in $\mathfrak{Q}$ was originated by Weinstein [W6], and was formulated in the language of Hilbert space by Aronszajn and Weinstein [AW1].

Occasionally, we shall also consider operators of the type

$$QAQu = \lambda u, \tag{3}$$

where $Q = I - P$ is an orthogonal projection operator onto a subspace $\mathfrak{Q}$. Here, $\mathfrak{Q}$ has finite codimension in $\mathfrak{H}$. We call QAQ the *projection of A to* $\mathfrak{Q}$. Much later than (2), several authors considered (3) without realizing the connection with (2) (see Sz.-Nagy [29], Halmos [H2], and Davis [D1]). It may be surmised that for this reason they did not discuss the spectrum.

Let us also remark that while the projection of A to $\mathfrak{Q}$ and the part of A in $\mathfrak{Q}$ coincide on $\mathfrak{Q}$, they are not completely equivalent. In fact, we shall see that the operator QAQ is an operator on $\mathfrak{H}$ (or dense subspace of $\mathfrak{H}$) while the *part of A in* $\mathfrak{Q}$ is strictly speaking only an operator on $\mathfrak{Q}$ (or a dense subspace of $\mathfrak{Q}$). Every eigenvalue of (2) is an eigenvector of (3) having the same corresponding eigenvalue. On the other hand, every nonzero $p \in \mathfrak{P}$ is obviously an eigenvector of QAQ corresponding to the eigenvalue zero, which will be called a *trivial* eigenvalue. However, such p's are clearly not eigenvectors of (2), since $Pp \neq 0$. Moreover, every eigenvector u of (3) such that $Pu = 0$ is an eigenvector of (2), having the same corresponding eigenvalue. These eigenvalues of (3) are called *nontrivial.*

The projection of A is used in the formulation of some theoretical questions such as the existence of the minimum for unbounded operators (Section 3), but is *not indispensible* except in the irreducible case, where it actually causes some difficulties (Section 9.4). On the other hand, let us emphasize the fact that to actually determine the eigenvalues we use the part of A in $\mathfrak{Q}$.

Limiting ourselves for the moment to a compact operator A, let us now show that $\lambda(p_1, p_2, \ldots, p_{n-1})$ is the first eigenvalue of (2), or equivalently, the first *nontrivial* eigenvalue of (3). For simplicity, put $\lambda(p_1, p_2, \ldots, p_{n-1}) = \tilde{\lambda}$ and let $\tilde{v}$ be a minimizing vector, that is, $\tilde{\lambda} = R(\tilde{v})$ and $\tilde{v} = Q\tilde{v}$. Using the very same reasoning as we did in the proof of Theorem 1.3.2, we have

$$AQ\tilde{v} - \tilde{\lambda}\tilde{v} = p \tag{4}$$

for some $p \in \mathfrak{P}$. Operating on both sides of (4) with Q yields

$$QAQ\tilde{v} - \tilde{\lambda}\tilde{v} = 0,$$

which proves that $\tilde{\lambda}$ is an eigenvalue of QAQ. Suppose that there is a smaller nontrivial eigenvalue $\tilde{\mu}$ of QAQ having eigenvector $\tilde{u}$. Then, it would follow that

$$\begin{aligned}\tilde{\lambda} &= \min_{\substack{u \in \mathfrak{D} \\ u = Qu}} R(u) \leq R(\tilde{u}) = (A\tilde{u}, \tilde{u})/(\tilde{u}, \tilde{u}) \\ &= (AQ\tilde{u}, Q\tilde{u})/(\tilde{u}, \tilde{u}) = (QAQ\tilde{u}, \tilde{u})/(\tilde{u}, \tilde{u}) = \tilde{\mu} < \tilde{\lambda},\end{aligned}$$

which is a contradiction. Therefore, $\tilde{\lambda}$ is the smallest nontrivial eigenvalue of QAQ.

Obviously, the value of $\lambda(p_1, p_2, \ldots, p_{n-1})$ depends only on the subspace spanned by $p_1, p_2, \ldots, p_{n-1}$ and not on the particular choice of vectors, which may or may not be independent. It is remarkable that in the classical formulation of this principle (1), the possibility of λ_n being a multiple eigenvalue does not play a part. This possibility can be handled by the following alternate formulation.

Theorem 1. *For a fixed n* ($n = 2, 3, \ldots$) *again let*

$$M = \max\{j \mid \lambda_j = \lambda_n\} \tag{5}$$

and let $\mathcal{T}_n$ *be the set of all subspaces* $\mathfrak{P}$ *of* $\mathfrak{H}$ *such that* $\dim \mathfrak{P} \leq M - 1$. *Then, the eigenvalues of A are given by the equation*

$$\lambda_n = \max_{\mathfrak{P} \in \mathcal{T}_n} \min_{\substack{u \in \mathfrak{D} \\ u \perp \mathfrak{P}}} R(u). \tag{6}$$

PROOF. For any $\mathfrak{P} \in \mathcal{T}_n$, let $r = \dim \mathfrak{P}$ and let $p_1, p_2, \ldots, p_r$ be any basis for $\mathfrak{P}$. We may choose $p_{r+1} = p_{r+2} = \cdots = p_{M-1} = 0$. Then we have

$$\min_{\substack{u \in \mathfrak{D} \\ u \perp \mathfrak{P}}} R(u) = \min_{\substack{u \in \mathfrak{D} \\ (u, p_i) = 0 \\ i = 1, 2, \ldots, r}} R(u) = \lambda(p_1, p_2, \ldots, p_r, p_{r+1}, \ldots, p_{M-1}) \leq \lambda_M = \lambda_n.$$

However, for the particular choice $\mathfrak{P} = \mathrm{sp}\{u_1, u_2, \ldots, u_{M-1}\}$, the classical characterization (1.3.2) yields the equality

$$\min_{\substack{u \in \mathfrak{D} \\ u \perp \mathfrak{P}}} R(u) = \lambda_M = \lambda_n,$$

which proves (6).

3. The Existence of Minima for Semibounded Operators

We now consider the minimum problem (1.1) for operators of class $\mathcal{S}$. As the existence of this minimum is not obvious, doubts have been expressed about this question, for instance, by Friedrichs [7, p. 127], Dunford and Schwartz [4, p. 1543], and also in an oral remark by Fichera [W18]. In particular, Friedrichs [7, p. 127] stated that he did not intend to prove that an actual minimum is assumed.† In this section, we fill this gap by proving, as in [S8], that the minimum (1.1) exists for operators of class $\mathcal{S}$. This result will be used in Chapters 4 and 7, but may be omitted by those who do not question the existence of the minimum.

Our proof of the proposition

$$\lambda(p_1, p_2, \ldots, p_{n-1}) = \min_{\substack{u \in \mathfrak{D} \\ (u, p_i) = 0 \\ i = 1, 2, \ldots, n-1}} R(u)$$

for operators of class $\mathcal{S}$ is motivated by the fact that, in the case of compact operators (see Section 2), the quantity $\lambda(p_1, p_2, \ldots, p_{n-1})$ is the first non-trivial eigenvalue of the problem

$$QAQu = \lambda u, \tag{1}$$

† In his text, Friedrichs [7] considers operators T such that $-T \in \mathcal{S}$.

where Q is the orthogonal projection onto $\mathfrak{P}^\perp$, the orthogonal complement of $\mathfrak{P} = \mathrm{sp}\{p_1, p_2, \ldots, p_{n-1}\}$. If we now let $A \in \mathscr{S}$, we see that QAQ is an operator of the type considered in Theorem A.2, and, therefore, it is selfadjoint. As mentioned before, every $p \in \mathfrak{P}$ $(p \neq 0)$ is an eigenvector of QAQ corresponding to the eigenvalue $\lambda = 0$. In the proof which follows, we show that the quantity

$$\mu(\mathfrak{P}) = \inf_{\substack{u \in \mathfrak{D} \\ Pu = 0}} R(u)$$

is the first *nontrivial* eigenvalue of QAQ and that the eigenvector corresponding to this nontrivial eigenvalue will be a vector for which the minimum $\lambda(p_1, p_2, \ldots, p_{n-1})$ is attained.

Theorem 1. *If $A \in \mathscr{S}$ has at least n isolated eigenvalues at the beginning of its spectrum, then the quantity*

$$\lambda(p_1, p_2, \ldots, p_{n-1}) = \min_{\substack{u \in \mathfrak{D} \\ (u, p_i) = 0 \\ i = 1, 2, \ldots, n-1}} R(u)$$

exists, is the first nontrivial eigenvalue of QAQ, namely $\lambda_1^{(n-1)}$, and satisfies the inequality

$$\lambda(p_1, p_2, \ldots, p_{n-1}) = \lambda_1^{(n-1)} \leq \lambda_n.$$

PROOF. Using the *test function* u_0 defined in the proof of Weyl's inequality, we see that

$$\mu(\mathfrak{P}) = \inf_{\substack{u \in \mathfrak{D} \\ Pu = 0}} R(u) \leq R(u_0) \leq \lambda_n. \tag{2}$$

Since λ_n is isolated, there exists a point λ_0 such that $(\lambda_n, \lambda_0]$ is in the resolvent set of A. Let $A_0 = A - \lambda_0 I$. We denote by E the projection operator onto $\mathfrak{U}_M = \mathrm{sp}\{u_1, u_2, \ldots, u_M\}$ where M is defined by (2.4.3). Observe that $E = E^2$ is the spectral projector E_λ for any $\lambda \in (\lambda_n, \lambda_0]$ and commutes with A and A_0. Therefore, we can write

$$A_0 = EA_0E + (I - E)A_0(I - E).$$

Let $K = EA_0E$ and $L = (I - E)A_0(I - E)$. Then, the operator K is compact (indeed, finite rank) and selfadjoint. Since A_0 is selfadjoint and $E\mathfrak{H}$ is finite-dimensional, we can apply Theorem A.2 to the operator L so that L is selfadjoint. Since $E = E_{\lambda_0}$, we see that K is nonpositive and L is nonnegative. Now, consider the operator

$$QA_0Q = QKQ + QLQ.$$

Note that the domain of QA_0Q consists of all $u \in \mathfrak{H}$ such that $Qu \in \mathfrak{D}$. It follows immediately that QKQ is compact, selfadjoint, and nonpositive. Since $\mathfrak{P}$ is finite-dimensional, we can apply Theorem A.2 to both QA_0Q and QLQ. Therefore, QA_0Q is selfadjoint and QLQ is selfadjoint and nonnegative. By construction, the domain of L is the same as the domain of A_0, which is of course $\mathfrak{D}$, the domain of A. Therefore, the domain of QLQ is $\mathfrak{P} \oplus \mathfrak{D} \cap \mathfrak{P}^{\perp}$. This implies

$$0 \leq \inf_{Qu \in \mathfrak{D}} [(QLQu, u)/(u, u)], \tag{3}$$

which means that the spectrum of QLQ has only nonnegative points. However, a result of Weyl (see Theorem A.10) states that the essential spectrum of a selfadjoint operator is unchanged by the addition of a compact selfadjoint operator. Thus, we see that every point of the essential spectrum of QA_0Q is nonnegative. On the other hand, inequality (2) implies

$$\inf_{\substack{u \in \mathfrak{D} \\ Pu = 0}} [(A_0 u, u)/(u, u)] < 0,$$

which is equivalent to

$$\inf[(QA_0Qu, u)/(Qu, Qu)] < 0.$$

Therefore, there exists $u = Qu \in \mathfrak{D}$ such that $(QA_0Qu, u)/(u, u) < 0$, which implies

$$\mu_* = \inf[(QA_0Qu, u)/(u, u)] < 0. \tag{4}$$

Equation (4) defines the lowest point of the spectrum of QA_0Q. Since $\mu_* < 0$, μ_* cannot be in the essential spectrum of QA_0Q, and therefore μ_* is the first isolated eigenvalue of QA_0Q. Furthermore, Eq. (4) can now be written as

$$\mu_* = \min_{Qu \in \mathfrak{D}} [(QA_0Qu, u)/(u, u)] < 0.$$

Let u_* be an eigenvector corresponding to μ_*. Then, we have

$$QA_0Qu_* = \mu_* u_*, \tag{5}$$

which means

$$Qu_* = u_*. \tag{6}$$

Using the fact that $A = A_0 + \lambda_0 I$ and Eqs. (6) and (5), we obtain

$$QAQu_* = \mu_* u_* + \lambda_0 u_*. \tag{7}$$

Equations (5) and (7) yield

$$(QA_0Qu_*, u_*)/(u_*, u_*) = \mu_*$$

and

$$(Au_*, u_*)/(u_*, u_*) = (QAQu_*, u_*)/(u_*, u_*) = \mu_* + \lambda_0 .$$

We can now write

$$\begin{aligned} \mu_* &= \frac{(QA_0Qu_*, u_*)}{(u_*, u_*)} = \min_{Qu \in \mathfrak{D}} \frac{(QA_0Qu, u)}{(u, u)} \\ &\le \inf_{\substack{u \in \mathfrak{D} \\ Pu=0}} \frac{(QA_0Qu, u)}{(u, u)} = \inf_{\substack{u \in \mathfrak{D} \\ Pu=0}} \frac{(Au, u) - \lambda_0(u, u)}{(u, u)} \\ &= \inf_{\substack{u \in \mathfrak{D} \\ Pu=0}} \frac{(Au, u)}{(u, u)} - \lambda_0 \le \frac{(Au_*, u_*)}{(u_*, u_*)} - \lambda_0 \\ &= \mu_* . \end{aligned} \tag{8}$$

Therefore, we must have equality throughout (8). In particular, we have

$$\mu(P) = \inf_{\substack{u \in \mathfrak{D} \\ Pu=0}} \frac{(Au, u)}{(u, u)} = \frac{(Au_*, u_*)}{(u_*, u_*)}$$

and therefore

$$\lambda(p_1, p_2, \ldots, p_{n-1}) = \min_{\substack{u \in \mathfrak{D} \\ Pu=0}} R(u) = (Au_*, u_*)/(u_*, u_*),$$

which shows that the minimum exists.

REMARK. Let us note that by using a *finite* number of orthogonality conditions we *cannot* obtain a minimum that is above the point λ_∞. In fact, the subspace $E_{\lambda_\infty}\mathfrak{H}$ is infinite-dimensional, so that for any n there always exists $u_0 \in E_{\lambda_\infty}\mathfrak{H}$ $(u_0 \neq 0)$ satisfying $(u_0, p_i) = 0$ $(i = 1, 2, \ldots, n-1)$, and therefore

$$\min R(u) \le R(u_0) \le \lambda_\infty .$$

4. Comparison of the Minimum–Maximum and Maximum–Minimum Principles

In view of the formal similarities and analogous proofs of the minimum–maximum and maximum–minimum principles, it must be emphasized here that these are two *fundamentally different principles* and not just a trivial interchange effected by reordering the eigenvalues. By ordering (as always) the eigenvalues in increasing sequence, the two principles are connected with *upper* and *lower* bounds, respectively. Curiously, the two principles have been confused with each other in numerous places in the literature, beginning with [3, p. 47].

At first glance, a difference between the minimum–maximum principle and the maximum–minimum principle appears to be that in the latter we have a variational problem on a (generally) infinite-dimensional subspace $\mathfrak{Q}$ of $\mathfrak{H}$ which is orthogonal to an arbitrary, finite-dimensional subspace $\mathfrak{P}$ of $\mathfrak{H}$, while in the former we have a variational problem on a finite-dimensional subspace $\mathfrak{V}$ of the domain $\mathfrak{D}$. However, one should not lose sight of the fact that both principles are based on variational problems

TABLE I

	Minimum-maximum	Maximum-minimum
Original inequalities due to	Poincaré (1890)	Weyl (1911)
Criteria for equalities due to	Stenger (1966)	Weinstein (1962) (and new max-min theory)
Arbitrary vectors (assume here independent)	$v_1, v_2, \ldots, v_r \in \mathfrak{D}$	$p_1, p_2, \ldots, p_{r-1} \in \mathfrak{H}$
Finite-dimensional subspace	$\mathfrak{V} = \operatorname{sp}(v_1, v_2, \ldots, v_r)$	$\mathfrak{P} = \operatorname{sp}(p_1, p_2, \ldots, p_{r-1})$
Restriction of dimensions	$m \leq r \leq M$	$m-1 \leq r-1 \leq M-1$
Associated projection	V onto $\mathfrak{V}$	Q onto $\mathfrak{P}^\perp$
Eigenvalue problem	$VAVu = \lambda u$	$QAQu = \lambda u$ or $\begin{Bmatrix} QAu = \lambda u \\ Qu = u \end{Bmatrix}$
Nontrivial eigenvalues	$\Lambda_1 \leq \Lambda_2 \leq \cdots \leq \Lambda_r$	$\lambda_1^{(r-1)} \leq \lambda_2^{(r-1)} \leq \cdots$
Eigenvectors	$w_1, w_2, \ldots, w_r$	$u_1^{(r-1)}, u_2^{(r-1)}, \ldots$
Eigenvalue inequalities	$\lambda_1 \leq \Lambda_1, \ldots, \lambda_r \leq \Lambda_r$	$\lambda_1^{(r-1)} \leq \lambda_r, \lambda_2^{(r-1)} \leq \lambda_{r+1}, \ldots$
Variational problem	$\Lambda_r = \max\limits_{u \in \mathfrak{V}} R(u)$	$\lambda_1^{(r-1)} = \min\limits_{u \in \mathfrak{P}^\perp} R(u)$
Variational characterization	$\lambda_n = \min\limits_{\mathfrak{V}} \max\limits_{u \in \mathfrak{V}} R(u)$	$\lambda_n = \max\limits_{\mathfrak{P}} \min\limits_{u \in \mathfrak{P}^\perp} R(u)$
Determinant giving necessary and sufficient conditions for equality	$\det\{([A - \lambda I]v_i, v_k)\}$ $i, k = 1, 2, \ldots, r$	$W_{r-1}(\lambda) = \det\{([A - \lambda I]^{-1}p_i, p_k)\}$ $i, k = 1, 2, \ldots, r-1$
Point of evaluation	$\lambda_n + \varepsilon; \varepsilon \geq 0$	$\lambda_n - \varepsilon; \varepsilon > 0$

on (generally) infinite-dimensional spaces [see the spectral formula (1.3.3) and the classical characterization (1.3.1)]. For compact operators, no additional theorems for infinite-dimensional spaces are required. For $\mathscr{S}$ operators in the infinite-dimensional case, we must investigate the (generally) unbounded operator QAQ, or the part of A in $\mathfrak{Q}$, which parallels the fact that lower bounds are intrinsically more difficult than upper bounds. Nevertheless, the modern ramifications of intermediate problems in several important cases succeeded in reducing the determination of lower bounds to a matrix problem, as was already the case of upper bounds.

Let us note that prior to the proof of the existence of the minimum in the maximum–minimum principle (Section 3), there was a corresponding sup-inf principle which would be sufficient neither for the theory of intermediate problems (Chapter 4) nor the new maximum–minimum theory (Chapter 7). Of course, in the minimum–maximum principle the use of inf-sup is superfluous, but has been done, even incorrectly, as was shown in [S8].

For the convenience of the reader we give in Table I some of the results of Chapters 2, 3, 4 and 7 which show the remarkable parallel between the two principles. While several results can be derived from either characterization, it should be noted that some results have only been obtained from the *maximum–minimum principle.* Among these is a contribution to the important theory of finite-rank perturbations (Section 9.2) and an inequality of Aronszajn's (Section 8.2). On the other hand, a recent inequality (Section 8.4) has been obtained thus far only by the *minimum–maximum principle.* In some cases, both characterizations are used simultaneously (Lemma 8.2.2).

5. Finite-Dimensional Spaces

Lax [L1] observed (without proof) that, if $\mathfrak{H}$ is a finite-dimensional space, the *maximum–minimum principle* and the *minimum–maximum principle* are in a certain sense equivalent. Essentially this is due to the fact that in this case the orthogonal complement of a finite-dimensional space is itself finite-dimensional.

Let us now give a proof of Lax's statement.

Theorem 1. *If $\mathfrak{H}$ is finite-dimensional, then the maximum–minimum principle* (2.1) *and the minimum–maximum principle* (2.2.3) *are equivalent.*

Proof. Let $\mathfrak{H}$ be N-dimensional and let A have the eigenvalues

$\lambda_1 \le \lambda_2 \le \cdots \le \lambda_N$. Let us also enumerate these eigenvalues in a descending sequence, namely

$$\mu_1 = \lambda_N, \quad \mu_2 = \lambda_{N-1}, \ldots, \quad \mu_{N+1-k} = \lambda_k, \ldots, \quad \mu_N = \lambda_1.$$

We define

$$m(\lambda_k) = \min\{i \mid \lambda_i = \lambda_k\}$$
$$m(\mu_k) = \min\{i \mid \mu_i = \mu_k\}$$
$$M(\lambda_k) = \max\{i \mid \lambda_i = \lambda_k\}$$
$$M(\mu_k) = \max\{i \mid \mu_i = \mu_k\}$$

so that we have (in general)

$$\lambda_{m(\lambda_k)-1} < \lambda_{m(\lambda_k)} = \lambda_{m(\lambda_k)+1} = \cdots = \lambda_k = \cdots = \lambda_{M(\lambda_k)} < \lambda_{M(\lambda_k)+1}$$

and

$$\mu_{m(\mu_k)-1} > \mu_{m(\mu_k)} = \mu_{m(\mu_k)+1} = \cdots = \mu_k = \cdots = \mu_{M(\mu_k)} > \mu_{M(\mu_k)+1}.$$

For a given index k ($k = 1, 2, \ldots, N$), we see that

$$m(\lambda_k) = \min\{i \mid \lambda_i = \lambda_k\}$$
$$= N + 1 - \max\{i \mid \mu_i = \mu_{N+1-k} = \lambda_k\} = N + 1 - M(\mu_{N+1-k}) \tag{1}$$

and

$$M(\lambda_k) = \max\{i \mid \lambda_i = \lambda_k\} = N + 1 - m(\mu_{N+1-k}). \tag{2}$$

We define the following classes of subspaces:

$$\mathscr{S}_k = \{\mathfrak{V} \subset \mathfrak{H} \mid m(\lambda_k) \le \dim \mathfrak{V}\}$$
$$\mathscr{T}_k = \{\mathfrak{P} \subset \mathfrak{H} \mid \dim \mathfrak{P} \le M(\lambda_k) - 1\}$$
$$\mathscr{S}_k' = \{\mathfrak{V}' \subset \mathfrak{H} \mid m(\mu_{N+1-k}) \le \dim \mathfrak{V}'\}$$
$$\mathscr{T}_k' = \{\mathfrak{P}' \subset \mathfrak{H} \mid \dim \mathfrak{P}' \le M(\mu_{N+1-k}) - 1\}.$$

Then, the variational principles (2.1) and (2.2.3) yield the equations

$$\lambda_k = \max_{\mathfrak{P} \in \mathscr{T}_k} \min_{u \perp \mathfrak{P}} R(u) \tag{3}$$

and

$$\lambda_k = \min_{\mathfrak{V} \in \mathscr{S}_k} \max_{u \in \mathfrak{V}} R(u). \tag{4}$$

For descending sequences, the roles of minimum and maximum are reversed, so that we have, for $\mu_{N+1-k} = \lambda_k$,

$$\mu_{N+1-k} = \min_{\mathfrak{P}' \in \mathscr{T}_k'} \max_{u \perp \mathfrak{P}'} R(u) \tag{5}$$

and

$$\mu_{N+1-k} = \max_{\mathfrak{V}' \in \mathscr{S}_k'} \min_{u \in \mathfrak{V}'} R(u). \tag{6}$$

If $\mathfrak{P}$ satisfies the inequality $\dim \mathfrak{P} + 1 \leq M(\lambda_k)$, then the space $\mathfrak{V}' = \mathfrak{P}^\perp$ satisfies the inequality

$$N + 1 - M(\lambda_k) \leq \dim \mathfrak{V}'.$$

Applying Eq. (2), we see that $m(\mu_{N+1-k}) \leq \dim \mathfrak{V}'$ and therefore (3) and (6) are equivalent. Similarly, if $\mathfrak{V}$ satisfies the inequality $m(\lambda_k) \leq \dim \mathfrak{V}$, then from (1) we see that the space $\mathfrak{P}' = \mathfrak{V}^\perp$ satisfies the inequality $\dim \mathfrak{P}' + 1 \leq M(\mu_{N+1-k})$. Therefore (4) and (5) are indeed equivalent.

Let us note that we do not have equivalence if $\mathfrak{H}$ is infinite–dimensional, even in the case of an operator of finite rank.

6. The General Maxi-Mini-Max Principle

In the preceding sections, certain relationships between the maximum–minimum and minimum–maximum principles have been mentioned. We now give a unifying variational characterization of eigenvalues, recently obtained by Stenger [S8], which includes each of the above characterizations as a special case.

Theorem 1. *(The maxi-mini-max principle) For a given pair of indices i, j ($i = 1, 2, \ldots; j = 0, 1, 2, \ldots$), let $\mathfrak{P}_j$ denote any j-dimensional subspace of $\mathfrak{H}$ and let $\mathfrak{V}_i$ denote any i-dimensional subspace of $\mathfrak{D} \cap \mathfrak{P}_j^\perp$. Then, the eigenvalues of A are given by the equation*

$$\lambda_{i+j} = \max_{\mathfrak{P}_j} \min_{\mathfrak{V}_i} \max_{u \in \mathfrak{V}_i} R(u). \tag{1}$$

Proof. Since the case $j = 0$ follows from Theorem 2.2.1, it suffices to consider indices $j \geq 1$. Let $p_1, p_2, \ldots, p_j$ denote a basis for $\mathfrak{P}_j$. Again letting Q_j denote the projector onto $\mathfrak{P}_j^\perp$, we see from Theorem 3.3.1 that the first eigenvalue $\lambda_1^{(j)}$ of $Q_j A Q_j$ is equal to $\lambda(p_1, p_2, \ldots, p_j)$ and satisfies the inequality $\lambda_1^{(j)} \leq \lambda_{j+1}$. Let $u_1^{(j)}$ be the eigenfunction corresponding to $\lambda_1^{(j)}$, consider $\lambda(p_1, p_2, \ldots, p_j, u_1^{(j)})$, and apply Theorem 3.3.1 and the classical characterization (1.3.2) to $\lambda_2^{(j)}$. We then obtain

$$\lambda(p_1, p_2, \ldots, p_j, u_1^{(j)}) = \lambda_2^{(j)} \leq \lambda_{j+2}.$$

Depending upon the number of isolated eigenvalues at the beginning of the spectrum of A, we can repeatedly apply the above reasoning to obtain the sequence of inequalities

$$\lambda(p_1, p_2, \ldots, p_j, u_1^{(j)}, u_2^{(j)}, \ldots, u_{i-1}^{(j)}) = \lambda_i^{(j)} \leq \lambda_{i+j} \qquad (i = 1, 2, \ldots).$$

Applying Theorem 2.2.1 to the operator $Q_j A Q_j$, we see, moreover, that

$$\lambda_{i+j} \geq \lambda_i^{(j)} = \min_{\mathfrak{V}_i} \max_{u \in \mathfrak{V}_i} R(u). \tag{2}$$

However, the classical choice $\mathfrak{P}_j = \mathfrak{U}_j$ yields equality in (2). Therefore, the maximum λ_{i+j} is attained and we have (1).

SPECIAL CASES. When $i = 1$, this characterization (1) reduces to *the maximum–minimum principle*; when $j = 0$, it reduces to the *minimum–maximum principle.*

In fact, if we put $i = 1$ into (1), then we only have to consider one-dimensional subspaces $\mathfrak{V}_1 \subset \mathfrak{D} \cap \mathfrak{P}_j^{\perp}$. Clearly, the Rayleigh quotient $R(u) = (Au, u)/(u, u)$ has a fixed value for all $u \in \mathfrak{V}_1$ ($u \neq 0$). Therefore,

$$\min_{\mathfrak{V}_1} \max_{u \in \mathfrak{V}_1} R(u)$$

can be replaced by

$$\min_{u \in \mathfrak{D} \cap \mathfrak{P}_j^{\perp}} R(u).$$

Then (1) becomes

$$\lambda_{1+j} = \max_{\mathfrak{P}_j} \min_{u \in \mathfrak{D} \cap \mathfrak{P}_j^{\perp}} R(u),$$

which is the maximum–minimum principle (3.2.1).

On the other hand, if we put $j = 0$ into (1), we see that the maximum over $\mathfrak{P}_0$ is superfluous, so that (1) becomes

$$\lambda_i = \min_{\mathfrak{V}_i} \max_{u \in \mathfrak{V}_i} R(u),$$

which is the minimum–maximum principle (2.2.3).

REMARK. In the course of the above proof, we have obtained the following significant result. *If the operator $A \in \mathscr{S}$ has an infinite sequence of initial eigenvalues, then every operator $Q_j A Q_j$ also has an infinite sequence of initial eigenvalues. If A has only a finite number, say N, initial eigenvalues, then $Q_j A Q_j$ has at least $N - j$ initial eigenvalues.*

Chapter Four

Intermediate Problems of the First Type

1. Weinstein's General Scheme of Intermediate Problems

As already discussed in Chapter 2, the Rayleigh–Ritz method yields upper bounds for operators of class $\mathscr{S}$. In 1935–37, Weinstein [W6–9] introduced a method for obtaining lower bounds which is now called the *method of intermediate problems*. More precisely, he considered, in modern terminology, intermediate problems of the first type. Later, other types of intermediate problems were considered, among which one of the most important is the second type of intermediate problems introduced by Aronszajn [A5].

Let us emphasize that the general scheme of the intermediate problems is the same regardless of type, so that, despite the differences in the numerical applications, it is nearly sufficient to develop the theory for the first type of problems.

It should be noted that the first type of intermediate problems has been a powerful instrument in its application to the numerical computation of lower bounds and has also been fruitful in the investigation of many general questions of a purely theoretical nature. The second type has also been successful in yielding numerical bounds for eigenvalues in many important problems. Moreover, the methods of solution themselves, which are patterned exactly after the first type, surprisingly led to the first determination of the spectrum of finite-rank perturbations (see Chapters 5 and 9).

The scheme of intermediate problems is the following: Given an eigenvalue problem for an operator of type $\mathscr{S}$, the first step is to find a *base problem*, namely an eigenvalue problem which has certain important properties to be described in detail later. The most outstanding property is that the eigenvalues of the base problem are all less than or equal to the corresponding eigenvalues of the given problem. The base problem will be our base of operation and is basic to the theory. This terminology will be used throughout. The *intermediate problems* are a sequence (often infinite) of eigenvalue problems which link the base problem to the given problems in such a way as to yield computable eigenvalues which are between those of the base and given problems. (Some authors call the base problem an "auxillary problem" and the intermediate operators "comparison operators." However, the terms *base problem* and *intermediate problems* are more descriptive.)

Let us note at once that the intermediate problems are not a part of the theory of perturbations of linear operators because we have here an unknown eigenvalue problem, while the theory of perturbations deals with the changes of the spectrum of a known operator. There are no base problems or intermediate problems in perturbation theory.

2. The Basic Principles of Intermediate Problems of the First Type

These problems were introduced and solved in terms of differential equations and boundary conditions for clamped plates (see Section 12). To facilitate our presentation of the theoretical part, we begin with the general scheme for operators in Hilbert space.

Let $\mathfrak{P}$ be a closed (usually infinite-dimensional) proper subspace of $\mathfrak{H}$ and let $\mathfrak{Q}$ be its orthogonal complement. Denote by P and Q the orthogonal projection operators onto $\mathfrak{P}$ and $\mathfrak{Q}$, respectively. We consider eigenvalue problems of the type

$$Au - PAu = \lambda u, \qquad Pu = 0, \tag{1}$$

which can also be written as

$$QAu = \lambda u, \qquad u = Qu. \tag{2}$$

We assume that A and QA are of class $\mathscr{S}$ in their respective domains of definition, namely on $\mathfrak{D}(A)$ and $\mathfrak{D}(A) \cap \mathfrak{Q}$. This assumption is satisfied not only for compact operators but for any $\mathscr{S}$ operator as well, by taking a suitable projection P, for instance, $P = E_\lambda$, $\lambda \in \sigma_e(A)$.

A useful way to consider (1) or (2) is to write the eigenvalue equation

$$QAQu = \lambda u, \tag{3}$$

where QAQ is defined for all u in $\mathfrak{P} \oplus \mathfrak{D}(A) \cap \mathfrak{Q}$, which for a compact A would be all of $\mathfrak{H}$. Let us point out that Eq. (3) is equivalent to (1) or (2) in the same sense as in the analogous situation discussed in Section 3.2.

For theoretical considerations, we assume that the spectral resolution of A is known. However, contrary to some opinions, e.g. [22, Russian ed.], in many applications it is sufficient to know *only a finite number of eigenvalues* $\lambda_1^{(0)} \leq \lambda_2^{(0)} \leq \cdots$ at the beginning of the spectrum of A and the corresponding eigenvectors $u_1^{(0)}, u_2^{(0)}, \ldots$ (see, for instance, Section 10). We call A the *base operator* and the corresponding eigenvalue problem the *base problem.*

Generally, the nontrivial eigenvalues $\lambda_1^{(\infty)} \leq \lambda_2^{(\infty)} \leq \cdots$ of the *given operator* QAQ cannot be determined explicitly. The Rayleigh–Ritz method (see Section 2.3) provides upper, but not lower, bounds for the nontrivial eigenvalues. By the principle of monotonicity (Section 2.5), the eigenvalues of A are not greater than the corresponding eigenvalues of QAQ. Therefore, the values $\lambda_1^{(0)}, \lambda_2^{(0)}, \ldots$ are lower bounds for $\lambda_1^{(\infty)}, \lambda_2^{(\infty)}, \ldots$. However, these lower bounds are usually very rough, as will be seen in applications.

In order to obtain improved lower bounds, we introduce a sequence of linearly independent vectors $p_1, p_2, \ldots$ in the subspace $\mathfrak{P}$. In the original problem of the plate, the vectors p_i were potential (or harmonic) functions, which explains the notation in our general case. We denote by $\mathfrak{P}_n$ the subspace generated by $p_1, p_2, \ldots, p_n$, and by P_n the orthogonal projection operator onto $\mathfrak{P}_n$. Let $Q_n = I - P_n$. Then the *intermediate problem of order* n, sometimes called the nth intermediate problem, is defined by the equations

$$Au - P_n Au = \lambda u, \qquad P_n u = 0. \tag{4}$$

Since $P_n Au \in \mathfrak{P}_n$, we can write

$$P_n Au = \alpha_1 p_1 + \alpha_2 p_2 + \cdots + \alpha_n p_n. \tag{5}$$

The scalars α_k are *not given* scalars but depend on the unknown u. To compute α_k, we take inner products with p_j $(j = 1, 2, \ldots, n)$ and obtain the system of equations

$$(P_n Au, p_j) = (Au, p_j) = \sum_{i=1}^{n} \alpha_i(p_i, p_j) \qquad (j = 1, 2, \ldots, n).$$

The α_k are then determined by Cramer's rule, since $\det\{(p_i, p_j)\}$ is just the Gram determinant of independent vectors and therefore not zero. If the p_k are orthonormal, we have the much simpler result $\alpha_k = (Au, p_k)$ $(k = 1, 2, \ldots, n)$. Let us note that if the condition $P_n u = 0$ were removed, we would have the nonsymmetric operator $A - P_n A$. However, since $P_n u = 0$, we have, for such u, $Q_n u = u$ and we can again write instead of (4)

$$Q_n A Q_n u = \lambda u, \qquad Q_n u = u. \tag{6}$$

Since for any $u \in \mathfrak{H}$, $Q_n u$ satisfies $P_n Q_n u = 0$, the operator $Q_n A Q_n$ can be extended to $u \in \mathfrak{H}$ such that $Q_n u \in \mathfrak{D}(A)$, which in the compact case would be to all $u \in \mathfrak{H}$. Obviously, $Q_n A Q_n$ on the subspace $P_n u = 0$ coincides with problem (4), which is implied by the term extension.

If A is compact, it is evident that $Q_n A Q_n$ is also a compact operator. If A is a general operator in $\mathscr{S}$, then, as has been proved in Section 3.3, $Q_n A Q_n$ is still in $\mathscr{S}$. We denote by $\lambda_1^{(n)} \leq \lambda_2^{(n)} \leq \cdots$ the eigenvalues and by $u_1^{(n)}, u_2^{(n)}, \ldots$ the corresponding orthonormal eigenvectors of Eq. (4), and for consistency we use the notation $\lambda_i^{(0)}$ and $u_i^{(0)}$ for the *base problem* and $\lambda_i^{(\infty)}$ and $u_i^{(\infty)}$ for the *given problem.*

Again from the principle of monotonicity we have the inequalities

$$\lambda_j^{(0)} \leq \lambda_j^{(1)} \leq \lambda_j^{(2)} \leq \cdots \leq \lambda_j^{(n)} \leq \cdots \leq \lambda_j^{(\infty)} \qquad (j = 1, 2, \ldots).$$

Therefore, we see that the eigenvalues of intermediate problems provide a sequence of nondecreasing lower bounds for the eigenvalues of the given problem.

The main property of intermediate problems, which actually justifies the whole theory, is that the eigenvalues of intermediate problems can be determined explicitly from the known spectrum of the base problem.

An eigenvalue $\lambda_j^{(n)}$ *of problem* (4) *is called persistent if it is an eigenvalue of the base problem*, that is, $\lambda_j^{(n)} = \lambda_i^{(0)}$ for some index i. By the principle of monotonicity we have $\lambda_i^{(0)} \leq \lambda_i^{(n)}$, so that

$$\begin{aligned} \max\{i \mid \lambda_i^{(n)} = \lambda_j^{(n)}\} &= \max\{i \mid \lambda_i^{(n)} \leq \lambda_j^{(n)}\} \\ &\leq \max\{i \mid \lambda_i^{(0)} \leq \lambda_j^{(n)}\} \\ &= \max\{i \mid \lambda_i^{(0)} = \lambda_j^{(n)}\}. \end{aligned}$$

In other words, in the sequences of eigenvalues of intermediate problems the persistent eigenvalues can move only *to the left.* In spite of this fact, in the Russian translation [12a, p. 155] such values are called *conservative.* An eigenvalue $\lambda_j^{(n)}$ is called *nonpersistent* if it is not an eigenvalue of the base problem, that is, $\lambda_k^{(n)} \neq \lambda_i^{(0)}$ for $i = 1, 2, \ldots$. The appearance of persistent eigenvalues is an essential point in the theory of intermediate problems and occurs even in the simplest one-dimensional engineering problems, such as the buckling of a clamped beam (see [W27]).

We begin with the discussion of nonpersistent eigenvalues, which is somewhat easier.

3. Nonpersistent Eigenvalues. Weinstein's Determinant

We have shown in Section 3.3 that the spectrum of (2.4) begins with isolated eigenvalues. We now discuss two ways of determining nonpersistent eigenvalues, the first of which was already given in 1935.

In the first approach, we begin by writing Eq. (2.4) as

$$Au - \lambda u = \alpha_1 p_1 + \alpha_2 p_2 + \cdots + \alpha_n p_n, \tag{1}$$

$$(u, p_i) = 0 \qquad (i = 1, 2, 3, \ldots, n). \tag{2}$$

At first glance, (1) appears to be an inhomogeneous problem. However, for any basis, the α_i are linear functionals of u, so that (1) is indeed homogeneous in u. Since λ is not in the spectrum of A, we can apply the resolvent $R_\lambda = [A - \lambda I]^{-1}$ to both sides of (1) and we then have

$$u = \sum_{i=1}^{n} \alpha_i R_\lambda p_i.$$

Using the orthogonality conditions (2), we obtain the set of equations

$$0 = (u, p_k) = \sum_{i=1}^{n} \alpha_i (R_\lambda p_i, p_k) \qquad (k = 1, 2, \ldots, n). \tag{3}$$

In Eq. (3), not all the α_i are zero since this would mean that λ is a persistent eigenvalue, contrary to our hypothesis. Therefore, the system (3) has a solution if and only if

$$W(\lambda) = \det\{(R_\lambda p_i, p_k)\} = 0 \qquad (i, k = 1, 2, \ldots, n). \tag{4}$$

The function $W(\lambda)$ is called the *Weinstein determinant*.

Let us note here that if we have a compact operator A, say negative-semidefinite, then $R_\lambda p_i = \sum_{j=1}^{\infty} (\lambda_j - \lambda)^{-1} (p_i, u_j) u_j$, so that

$$W(\lambda) = \det\left\{\sum_{j=1}^{\infty} \frac{(p_i, u_j)(u_j, p_k)}{\lambda_j - \lambda}\right\} \qquad (i, k = 1, 2, \ldots, n) \tag{5}$$

Similarly, for $A \in \mathscr{S}$ we have

$$W(\lambda) = \det\left\{\sum_{j} \frac{(p_i, u_j)(u_j, p_k)}{\lambda_j - \lambda} + \int_{\lambda_\infty - 0}^{\infty} \frac{1}{\mu - \lambda}\, d(E_\mu p_i, p_k)\right\}$$

$$(i, k = 1, 2, \ldots, n). \tag{6}$$

We see from these representations (5) and (6) that the matrix of $W(\lambda)$ is always symmetric and that $W(\lambda)$ has a meromorphic character at the isolated eigenvalues. More on its general analytic character will be discussed later (see Section 8).

If λ is a nonpersistent root of $W(\lambda)$, then there exist linearly independent solutions

$$a^{(1)} = [\alpha_1^{(1)}, \alpha_2^{(1)}, \ldots, \alpha_n^{(1)}],$$

$$a^{(2)} = [\alpha_1^{(2)}, \alpha_2^{(2)}, \ldots, \alpha_n^{(2)}],$$

$$\vdots$$

$$a^{(\nu)} = [\alpha_1^{(\nu)}, \alpha_2^{(\nu)}, \ldots, \alpha_n^{(\nu)}]$$

of system (3), where ν denotes the nullity of the matrix $\{(R_\lambda p_i, p_k\}$. Here, the nullity, as usual, is defined as n minus the rank. For each $a^{(j)}$, there is a nontrivial solution $v^{(j)}$ of Eq. (1) given by

$$v^{(j)} = \sum_{i=1}^{n} \alpha_i^{(j)} R_\lambda p_i \qquad (j = 1, 2, \ldots, \nu).$$

These solutions are linearly independent. In fact, if we would have scalars $\beta^{(j)}$ such that $\sum_{j=1}^{\nu} \beta^{(j)} v^{(j)} = 0$, then it would follow that

$$\sum_{i=1}^{n} \sum_{j=1}^{\nu} \beta^{(j)} \alpha_i^{(j)} p_i = (A - \lambda I) \sum_{j=1}^{\nu} \beta^{(j)} v^{(j)} = 0.$$

Since the p_i are linearly independent, it means that the coefficients $\sum_{j=1}^{\nu} \beta^{(j)} \alpha_i^{(j)}$ are zero for $i = 1, 2, \ldots, n$. This implies that $\sum_{j=1}^{\nu} \beta^{(j)} a^{(j)} = 0$, and since the $a^{(j)}$ are linearly independent, we have $\beta^{(1)} = \beta^{(2)} = \cdots = \beta^{(\nu)} = 0$.

We can summarize this result as follows.

Theorem 1. *The value λ is a nonpersistent eigenvalue of Eqs.* (1) and (2) *if and only if λ is a* (*nonpersistent*) *root of the determinant* (4),

$$W(\lambda) = \det\{(R_\lambda p_i, p_k)\} = 0 \qquad (i, k = 1, 2, \ldots, n).$$

Moreover, the multiplicity of λ is equal to the nullity ν of the matrix

$$\{(R_\lambda p_i, p_k)\} \qquad (i, k = 1, 2, \ldots, n). \tag{7}$$

Let us mention that we actually used in the above proof the following simple general principle. If T is any invertible operator and $p_1, p_2, \ldots, p_n$ are independent vectors in $\mathfrak{D}(T)$, then $Tp_1, Tp_2, \ldots, Tp_n$ are also independent. In fact, if we would have $\sum_{i=1}^{n} \xi_i Tp_i = 0$, that is, $T(\sum_{i=1}^{n} \xi_i p_i) = 0$, then it would follow that $\sum_{i=1}^{n} \xi_i p_i = 0$, and, therefore, $\xi_1 = \xi_2 = \cdots = \xi_n = 0$.

We now give a second proof of Theorem 1 which is of considerable interest because it parallels an analogous proof for intermediate problems of the second type (see Section 5.2).

Assuming again that λ is not an eigenvalue of A, we consider the equation

$$Au - P_n Au = \lambda u, \tag{8}$$

or

$$Au - \lambda u = \alpha_1 p_1 + \alpha_2 p_2 + \cdots + \alpha_n p_n \tag{9}$$

where, as before, the α_i are linear functionals of the unknown u. Now, instead of using the orthogonality conditions (2), we consider (8) for all

$u \in \mathfrak{D}$. In this case, we actually have a (generally) nonsymmetric operator. Therefore, unlike in the first proof, we do not know here *a priori* that the spectrum is real and begins with isolated eigenvalues. Nevertheless, we can discuss the spectrum in the following way [W23].

As λ is not in the spectrum of A, we have the solution of (9) given by

$$u = R_\lambda\left[\sum_{i=1}^{n} \alpha_i p_i\right] = \sum_{i=1}^{n} \alpha_i R_\lambda p_i, \tag{10}$$

where of course

$$AR_\lambda p_i - \lambda R_\lambda p_i = p_i \qquad (i = 1, 2, \ldots, n). \tag{11}$$

We now use the fact that, for u satisfying (10), we have

$$P_n Au = \alpha_1 p_1 + \alpha_2 p_2 + \cdots + \alpha_n p_n. \tag{12}$$

Therefore, we have from (10) and (12)

$$P_n\left[\sum_{i=1}^{n} \alpha_i AR_\lambda p_i\right] - [\alpha_1 p_1 + \alpha_2 p_2 + \cdots + \alpha_n p_n] = 0.$$

In order to solve for the α_i, we take the scalar product with the p_k, thus obtaining

$$\sum_{i=1}^{n} \alpha_i[(AR_\lambda p_i, p_k) - (p_i, p_k)] = 0 \tag{13}$$

or equivalently

$$\sum_{i=1}^{n} \alpha_i([AR_\lambda - I]p_i, p_k) = 0 \tag{14}$$

for $k = 1, 2, \ldots, n$. Now, since $R_\lambda = [A - \lambda I]^{-1}$, we can write

$$AR_\lambda - I = (A - [A - \lambda I])R_\lambda = \lambda R_\lambda,$$

so that Eq. (14) becomes

$$\sum_{i=1}^{n} \alpha_i \lambda(R_\lambda p_i, p_k) = 0 \qquad (k = 1, 2, \ldots, n). \tag{15}$$

Nontrivial solutions of (15) exist if and only if

$$\lambda^n W(\lambda) = 0, \tag{16}$$

which, compared with (4), seemingly introduces an additional factor λ^n. We now show that this factor does not change the result stated in Theorem 1. For $\lambda \neq 0$, $W(\lambda)$ must be zero and we have the same condition as in the first proof. The fact that we have obtained the result without using explicitly the orthogonality conditions (2) can be explained in the following way. If $\lambda \neq 0$ and u is a solution of (1), then we can write

$$u = (1/\lambda)(Au - P_n Au). \tag{17}$$

If we now take P_n of both sides of (17), we have

$$P_n u = (1/\lambda)(P_n Au - P_n Au) = 0,$$

so that u *automatically satisfies the orthogonality conditions* (2), which were not used in the present proof as long as $\lambda \neq 0$.

Let us now consider the possibility of $\lambda = 0$ being a nonpersistent eigenvalue. Of course, this case cannot occur if A is a compact operator on an infinite-dimensional space, since zero is always in the spectrum of A. However, for an operator A of class $\mathcal{S}$, $\lambda = 0$ may be in the resolvent set for A, that is, A^{-1} exists and is bounded. This means $Au = 0$ only if $u = 0$. To show that the factor λ^n does not modify the theorem, let us suppose that $\lambda = 0$ is a nonpersistent eigenvalue of the intermediate problem (1). Then, we have

$$Au = \alpha_1 p_1 + \alpha_2 p_2 + \cdots + \alpha_n p_n$$

and therefore

$$u = \sum_{i=1}^{n} \alpha_i R_0 p_i,$$

where of course $R_0 = A^{-1}$. Now, using again the condition $P_n u = 0$, we obtain

$$0 = \sum_{i=1}^{n} \alpha_i (R_0 p_i, p_k) \qquad (k = 1, 2, \ldots, n),$$

which has nontrivial solutions if and only if $W(0) = 0$. Thus, we have the same result as has been formulated in the theorem, including the rule for multiplicity.

In order to elucidate the role of the factor λ^n in (16), we consider the *nonsymmetric* eigenvalue problem

$$Au - P_n Au = \lambda u, \tag{18}$$

where the condition $P_n u = 0$ is dropped regardless of whether λ is zero or nonzero.

Now, let us suppose that $\lambda = 0$ is not in the spectrum of A. Then, we have the following result.

Theorem 2. *If zero is not in the spectrum of A, then the equation*

$$Au - P_n Au = 0 \tag{19}$$

has exactly n linearly independent solutions. Moreover, if ν is the nullity of the matrix (7) *(for $\lambda = 0$), then exactly ν of the solutions satisfy the orthogonality conditions* (2).

PROOF. Taking $p_1, p_2, \ldots, p_n$ as any basis for P_n, the n solutions are given by

$$u_{0,i} = A^{-1}p_i \qquad (i = 1, 2, \ldots, n). \tag{20}$$

If $u_{0,n+1}$ were another independent solution of (19), we would have

$$Au_{0,n+1} = \sum_{i=1}^{n} \alpha_i p_i$$

for α_i not all zero, and therefore

$$u_{0,n+1} = \sum_{i=1}^{n} \alpha_i A^{-1}p_i = \sum_{i=1}^{n} \alpha_i u_{0,i},$$

which is a contradiction. We now consider the matrix

$$\{(R_0 p_i, p_k)\} = \{(A^{-1}p_i, p_k)\} \qquad (i, k = 1, 2, \ldots, n). \tag{21}$$

As in the first method for the determination of nonpersistent eigenvalues, there are ν solutions satisfying (10). If there were more than ν solutions satisfying (10), then the matrix (21) would have nullity at least $\nu + 1$, contrary to the assumption.

Let us note in passing that if $\lambda = 0$ is in the spectrum of A, for instance if A is compact and $\dim \mathfrak{H} = \infty$, a result similar to Theorem 2 could be given.

4. The Distinguished Choice

There are three methods for the determination of persistent eigenvalues in intermediate problems of the first type, all of which can be used also for the second type. We begin with the oldest method, due to Weinstein [W9], which leads to various ramifications. In problems of the second type, it was reintroduced in a modified form by Bazley [B3].

The essence of this method is to reduce the determination of a persistent eigenvalue to the problem of determining a nonpersistent eigenvalue. If λ_* is an eigenvalue of the base problem, we introduce first an intermediate problem in which λ_* does not appear as an eigenvalue and use this intermediate problem as a new base problem. Let λ_* have multiplicity μ and let $u^1, u^2, \ldots, u^\mu$ be an orthonormal basis for the eigenspace $\mathfrak{U}_*$ of λ_*. We now choose vectors $p^1, p^2, \ldots, p^\mu \in \mathfrak{P}$ in a special way, namely, so that

$$(p^i, u^k) = \delta_{ik} \qquad (i, k = 1, 2, \ldots, \mu). \tag{1}$$

The existence of such vectors will be discussed in the next section. These vectors define a subspace $\mathfrak{P}_\mu$ and we consider the corresponding intermediate problem of order μ. Since the intermediate problem only depends on the subspace and not on the basis chosen, condition (1) can be replaced by an equivalent condition

$$\det\{(p^i, u^k)\} \neq 0 \qquad (i, k = 1, 2, \ldots, \mu) \tag{2}$$

which is valid for any basis. It is obvious that (1) implies (2). On the other hand, using (2), we can find vectors

$$q^j = \sum_{i=1}^{\mu} \alpha_{ji} p^i \qquad (j = 1, 2, \ldots, \mu)$$

so that

$$(q^j, u^k) = \sum_{i=1}^{\mu} \alpha_{ji}(p^i, u^k) = \delta_{jk}.$$

Indeed, $\{\alpha_{ji}\}$ is just the matrix inverse to $\{(p^i, u^k)\}$. Such a *special selection* of the p^i has been called *Weinstein's distinguished choice.*

Let us now consider our intermediate problem of order μ, which by (3.1) can be written as

$$Au - \lambda_* u = \alpha_1 p^1 + \alpha_2 p^2 + \cdots \alpha_\mu p^\mu, \qquad P_\mu u = 0. \tag{3}$$

This problem does not admit the eigenvalue λ_*. In fact, as λ_* is an eigenvalue of the base problem, we know that a necessary and sufficient condition for (3) to have a solution is the Fredholm condition, that is, the right-hand side must be orthogonal to $\mathfrak{U}_*$. This yields a system of equations

$$\sum_{i=1}^{\mu} \alpha_i(p^i, u^k) = 0 \qquad (k = 1, 2, \ldots, \mu). \tag{4}$$

Since by (2) the determinant of the coefficients in (4) is not zero, the only solution is $\alpha_1 = \alpha_2 = \cdots = \alpha_\mu = 0$, which means $u = \gamma_1 u^1 + \gamma_2 u^2 + \cdots + \gamma_\mu u^\mu$. However, since $P_\mu u = 0$, it follows that $u = 0$, where we again use the fact that the determinant in (4) is not zero. This means that λ_* is not persistent.

We now use the intermediate problem (3) as a *new* base problem relative to which we construct an intermediate problem of order σ by introducing new independent vectors $p_{\mu+1}, p_{\mu+2}, \ldots, p_{\mu+\sigma}$. As the new base problem does not contain λ_* as an eigenvalue, the question of the persistency of λ_* would be eliminated if our only aim were to express the eigenvalues of subsequent intermediate problems in terms of the eigenelements of the new base problem. However, our purpose is to solve subsequent intermediate problems in terms of the original base problem. This is achieved in the following way, reminiscent of Huygens' principle (see also Section 8).

Obviously, an intermediate problem of order σ relative to the new base problem is an intermediate problem of order $\mu + \sigma$ relative to the original base problem, $Au = \lambda u$. It is given by the equations

$$Au - P_\mu Au - \lambda_* u = \alpha_{\mu+1} p_{\mu+1} + \cdots + \alpha_{\mu+\sigma} p_{\mu+\sigma}, \qquad P_{\mu+\sigma} u = 0. \quad (5)$$

This problem (5) may again admit λ_* as an eigenvalue, except of course in the case when $\lambda_* = \lambda_1^{(0)}$, because by the monotonicity principle the index of $\lambda_1^{(0)}$ in the new problem would have to be less than one, which is impossible.

To solve this problem, we proceed as follows. First of all, we redefine the basis of $\mathfrak{P}_{\mu+\sigma}$ in such a way that the vectors $p^1, p^2, \ldots, p^\mu$ are preserved and that the vectors $p_{\mu+1}, p_{\mu+2}, \ldots, p_{\mu+\sigma}$ are replaced by vectors $p^{\mu+1}$, $p^{\mu+2}, \ldots, p^{\mu+\sigma}$ each of which is orthogonal to $\mathfrak{U}_*$. This is done by defining

$$p^{\mu+i} = p_{\mu+i} - \sum_{j=1}^{\mu} \gamma_{ij} p^j \qquad (i = 1, 2, \ldots, \sigma).$$

Then, we have

$$(p^{\mu+i}, u^k) = (p_{\mu+i}, u^k) - \sum_{j=1}^{\mu} \gamma_{ij}(p^j, u^k) = 0 \qquad (i = 1, 2, \ldots, \sigma). \quad (6)$$

For a given i, the nonhomogeneous equation (6) can be solved since the determinant (2) is not zero.

We now rewrite our intermediate problem of order $\mu + \sigma$ in the form

$$Au - \lambda_* u = \beta_1 p^1 + \beta_2 p^2 + \cdots + \beta_\mu p^\mu + \beta_{\mu+1} p^{\mu+1} + \cdots + \beta_{\mu+\sigma} p^{\mu+\sigma}, \quad (7)$$

with β_i playing the role of α_i $(i = 1, 2, \ldots, \mu + \sigma)$. Necessary and sufficient conditions for a solution of (7) to exist are

$$\left(\sum_{i=1}^{\mu} \beta_i p^i + \sum_{i=1}^{\sigma} \beta_{\mu+i} p^{\mu+i}, u^k \right) = 0 \qquad (k = 1, 2, \ldots, \mu).$$

However, since we already have $(p^{\mu+i}, u^k) = 0$ $(i = 1, 2, \ldots, \sigma;\ k = 1, 2, \ldots, \mu)$ our condition reduces to

$$\sum_{i=1}^{\mu} \beta_i(p^i, u^k) = 0 \qquad (k = 1, 2, \ldots, \mu). \tag{8}$$

As we have a distinguished choice, it follows that $\beta_1 = \beta_2 = \cdots = \beta_\mu = 0$, so that Eq. (7) becomes

$$Au - \lambda_* u = \beta_{\mu+1} p^{\mu+1} + \cdots + \beta_{\mu+\sigma} p^{\mu+\sigma}. \tag{9}$$

Since the right-hand side of (9) is orthogonal to $\mathfrak{U}_*$, Eq. (9) has infinitely many solutions, which are given by

$$u = \sum_{i=1}^{\sigma} \beta_{\mu+i} R'_{\lambda_*} p^{\mu+i} + \gamma_1 u^1 + \gamma_2 u^2 + \cdots + \gamma_\mu u^\mu. \tag{10}$$

Let us explain here the meaning of R'_{λ_*}. The ordinary resolvent R_{λ_*} is not defined since $A - \lambda_* I$ is not one-to-one. However, if we consider $A - \lambda_* I$ on the subspace $\mathfrak{U}_*^{\perp}$ [actually $\mathfrak{U}_*^{\perp} \cap \mathfrak{D}(A)$], then $A - \lambda_* I$ is one-to-one and onto $\mathfrak{U}_*^{\perp}$. In fact, if we had $f_1, f_2 \in \mathfrak{U}_*^{\perp} \cap \mathfrak{D}(A)$ such that

$$Af_1 - \lambda_* f_1 = Af_2 - \lambda_* f_2,$$

then $f_1 - f_2$ would be in $\mathfrak{U}_*$ and therefore $f_1 = f_2$. For any g, we define $R'_{\lambda_*} g$ by

$$R'_{\lambda_*} g = \sum_{\lambda_j^{(0)} \neq \lambda_*} \frac{(g, u_j^{(0)})}{\lambda_j^{(0)} - \lambda_*} u_j^{(0)} + \int_{\lambda_\infty - 0}^{\infty} \frac{1}{\lambda - \lambda_*} \, dE_\lambda g. \tag{11}$$

Let us note the important fact that $R'_{\lambda_*} g$ is orthogonal to $\mathfrak{U}_*$. Moreover, if g itself is orthogonal to $\mathfrak{U}_*$, we have

$$(A - \lambda_* I) R'_{\lambda_*} g = g.$$

In order to have a nonvanishing solution of (9) which is orthogonal to $\mathfrak{P}_{\mu+\sigma}$, we must satisfy the equations

$$(u, p^k) = 0 \qquad (k = 1, 2, \ldots, \mu + \sigma). \tag{12}$$

Let us now show that $\beta_{\mu+1}, \beta_{\mu+2}, \ldots, \beta_{\mu+\sigma}$ cannot all be zero. If they were, we would have, by (10) and (12),

$$\sum_{i=1}^{\mu} \gamma_i(u^i, p^k) = 0 \qquad (k = 1, 2, \ldots, \mu).$$

As the choice is distinguished, all γ_i would also have to be zero, thereby giving only the trivial solution.

The orthogonality conditions (12) are divided into two parts in a natural way. First of all, in view of conditions (6), the orthogonality conditions (12) for $k = \mu + 1, \mu + 2, \ldots, \mu + \sigma$ contain only the unknowns $\beta_{\mu+1}$, $\beta_{\mu+2}, \ldots, \beta_{\mu+\sigma}$. Therefore, if λ_* is a persistent eigenvalue of (9), there has to be some nonzero coefficient $\beta_{\mu+i}$. This means that the determinant

$$W(\lambda_*) = \det\{(R'_{\lambda_*} p^{\mu+i}, p^{\mu+k})\} \qquad (i, k = 1, 2, \ldots, \sigma) \tag{13}$$

must be zero. Let us emphasize here the fact that $W(\lambda_*)$ is nothing else but a Weinstein determinant considered only for vectors p which are orthogonal to $\mathfrak{U}_*$. The independent solutions for the $\beta_{\mu+i}$ are determined by the nullity ν of the matrix in (13). For each of these solutions, the orthogonality conditions (12) for $k = 1, 2, \ldots, \mu$ are satisfied in view of (2) by a proper choice of $\gamma_1, \gamma_2, \ldots, \gamma_\mu$.

Let us now show that the multiplicity of λ_* in our intermediate problem of order $\mu + \sigma$, (5), is ν. We denote by

$$v^{(j)} = \sum_{i=1}^{\sigma} \beta_{\mu+i}^{(j)} R'_{\lambda_*} p^{\mu+i} + \sum_{i=1}^{\mu} \gamma_i^{(j)} u^i \qquad (j = 1, 2, \ldots, \nu) \tag{14}$$

the solutions of (9) that correspond to the ν independent solutions $b^{(j)} = [\beta_{\mu+1}^{(j)}, \beta_{\mu+2}^{(j)}, \ldots, \beta_{\mu+\sigma}^{(j)}]$ for the coefficients $\beta_{\mu+i}$. Suppose there exist $\alpha^{(1)}, \alpha^{(2)}, \ldots, \alpha^{(\nu)}$ not all zero such that $\sum_{j=1}^{\nu} \alpha^{(j)} v^{(j)} = 0$. Since the vectors $R'_{\lambda_*} p^{\mu+i}$ are all orthogonal to $\mathfrak{U}_*$, it follows from (14) that both

$$\sum_{j=1}^{\nu} \alpha^{(j)} \sum_{i=1}^{\sigma} \beta_{\mu+i}^{(j)} R'_{\lambda_*} p^{\mu+i} = 0 \tag{15}$$

and

$$\sum_{j=1}^{\nu} \alpha^{(j)} \sum_{i=1}^{\mu} \gamma_i^{(j)} u^{(j)} = 0.$$

Since the $p^{\mu+i}$ are independent and R'_{λ_*} is invertible in $\mathfrak{U}_*^{\perp}$, we have by the general principle (see Section 3) the fact that the $R'_{\lambda_*} p^{\mu+i}$ are independent. It follows, therefore, from (15) that

$$\sum_{j=1}^{\nu} \alpha^{(j)} \beta_{\mu+i}^{(j)} = 0 \qquad (i = 1, 2, \ldots, \sigma). \tag{16}$$

Since the $b^{(j)} = [\beta_{\mu+1}^{(j)}, \ldots, \beta_{\mu+\sigma}^{(j)}]$ are independent, we have $\alpha^{(1)} = \alpha^{(2)} = \cdots = \alpha^{(\nu)} = 0$.

From formula (11), it is seen that if λ_* reappears in the problem of order $\mu + \sigma$, the corresponding *new* eigenvector is never in $\mathfrak{U}_*$. In fact, $R'_{\lambda_*} p^{\mu+i}$ is always orthogonal to $\mathfrak{U}_*$.

Assuming as above that a given λ_*, say $\lambda_k^{(0)}$, admits a distinguished choice, we have the following theorems.

Theorem 1. *For a given index k, if $\lambda_k^{(0)}$, having multiplicity μ, admits a distinguished choice $p^1, p^2, \ldots, p^\mu$,* then *$\lambda_k^{(0)}$ is strictly less than $\lambda_k^{(\mu)}$, the corresponding eigenvalue of the intermediate problem* (3).

REMARK. It should be emphasized that the principle of monotonicity (Section 2.5) yields only the less sharp inequality $\lambda_k^{(0)} \leq \lambda_k^{(\mu)}$.

PROOF. Since we always have the inequality $\lambda_k^{(0)} \leq \lambda_k^{(\mu)}$, and by the results just derived, we know that $\lambda_k^{(0)}$ is not an eigenvalue of our intermediate problem (3), we have $\lambda_k^{(0)} < \lambda_k^{(\mu)}$.

Corollary 1. *If the subspace $\mathfrak{P}_{\mu+\sigma}$ defining an intermediate problem of order $\mu + \sigma$ contains the vectors $p^1, p^2, \ldots, p^\mu$ of Theorem* 1, *then $\lambda_k^{(0)}$ is strictly less than $\lambda_k^{(\mu+\sigma)}$.*

PROOF. The proof is obvious since $\lambda_k^{(\mu)} \leq \lambda_k^{(\mu+\sigma)}$.

Corollary 2. *Under the hypothesis of Theorem* 1, *we have the strict inequality $\lambda_k^{(0)} < \lambda_k^{(\infty)}$.*

PROOF. In fact, we have

$$\lambda_k^{(0)} < \lambda_k^{(\mu)} \leq \lambda_k^{(\mu+1)} \leq \cdots \leq \lambda_k^{(\infty)}.$$

5. Lower Bounds Using a Distinguished Choice

We now develop the rules for determining improved lower bounds for the eigenvalues of the given problem by using distinguished choices. This method will yield improved lower bounds for a finite but arbitrarily large number of eigenvalues. Let us note in passing that the Rayleigh–Ritz method, gives upper bounds also for a *finite* number of eigenvalues, *but improvement is not guaranteed*, as in our case for lower bounds (see Section 2.3).

Let us suppose we wish to find improved lower bounds for $\lambda_1^{(\infty)} \leq \lambda_2^{(\infty)} \leq \cdots \leq \lambda_n^{(\infty)}$, where n is any arbitrary but fixed index. In this section, we assume that distinguished choices exist for all $\lambda_i^{(0)}$, or for at least $\lambda_1^{(0)} \leq \lambda_2^{(0)} \leq \cdots \leq \lambda_n^{(0)}$. This assumption will be removed in the next section. We form a subspace $\mathfrak{P}_r \subset \mathfrak{P}$, which contains a distinguished choice for each $\lambda_i^{(0)}$ $(i = 1, 2, \ldots, n)$. Since distinguished choices belonging to different eigenvalues are not necessarily independent, the dimension r of $\mathfrak{P}_r$ is not greater than the sum of the multiplicities of the distinct eigenvalues $\lambda_i^{(0)} \leq \lambda_n^{(0)}$.

We first consider an intermediate problem of order r given by

$$Au - P_r Au = \lambda u, \qquad P_r u = 0. \tag{1}$$

Let $\tilde{\lambda}$ be the next point in the spectrum after $\lambda_n^{(0)}$. The point $\tilde{\lambda}$ is either the next higher eigenvalue of the base problem or the first limit point of the spectrum of A, for instance, if A is an operator of finite rank. We determine all nonpersistent eigenvalues (if any) that are less than $\tilde{\lambda}$. For the moment, we call such values $\rho_1 \le \rho_2 \le \cdots \le \rho_s$. Next, we determine separately all persistent eigenvalues (if any) that are less than $\tilde{\lambda}$, say $\sigma_1 \le \sigma_2 \le \cdots \le \sigma_t$, by the rules of Section 4. Then, we arrange the ρ_i and σ_k thus obtained in one ascending sequence. In this way, we have all eigenvalues $\lambda_1^{(r)} \le \lambda_2^{(r)} \le \cdots \le \lambda_q^{(r)}$ of (1) that are below $\tilde{\lambda}$. It is important to note that there may be no such eigenvalues, because of a wholesale improvement. Nonetheless, in all cases, we obtain improved lower bounds for $\lambda_1^{(\infty)} \le \lambda_2^{(\infty)} \le \cdots \le \lambda_n^{(\infty)}$ since for some q $(0 \le q \le n)$ we have the following inequalities:

$$\begin{aligned}
&\lambda_1^{(0)} < \lambda_1^{(r)} \le \lambda_1^{(\infty)}\\
&\lambda_2^{(0)} < \lambda_2^{(r)} \le \lambda_2^{(\infty)}\\
&\quad\vdots\\
&\lambda_q^{(0)} < \lambda_q^{(r)} \le \lambda_q^{(\infty)}\\
&\lambda_n^{(0)} < \tilde{\lambda} \le \lambda_{q+1}^{(\infty)} \le \lambda_{q+2}^{(\infty)} \le \cdots \le \lambda_n^{(\infty)}.
\end{aligned} \tag{2}$$

This property of definite improvement makes the method highly valuable in numerical applications. Formally, we have also found lower bounds for all eigenvalues of the given problem, since we already have the inequalities

$$\lambda_{n+i}^{(0)} \le \lambda_{n+i}^{(\infty)}, \qquad i = 1, 2, \ldots .$$

6. The Existence of Distinguished Choices

In the previous section, we have assumed the existence of a distinguished choice for each eigenvalue of the base problem. In such cases, by writing down the distinguished choice for each successive eigenvalue and suppressing those vectors that are dependent on the previous ones, we obtain a sequence $p_1, p_2, \ldots$ which we call a *distinguished sequence*. A distinguished sequence has the property that it contains a distinguished choice for each eigenvalue of the base problem. For examples of explicit constructions of distinguished sequences, see Sections 12 and 8.3.

We now consider intermediate problems in the general case when there is no distinguished sequence. We shall show that this case is reducible to the method described in the previous section.

We begin with the following theorem.

Theorem 1. *Let λ_* be an eigenvalue of the base problem having multiplicity μ and let $\mathfrak{U}_*$ be the corresponding eigenspace. A distinguished choice for λ_* exists if and only if no vector in $\mathfrak{U}_*$ is an eigenvector of the given problem* (2.1).

PROOF. First, let $p^1, p^2, \ldots, p^\mu$ be a distinguished choice for λ_*. Then, by Eq. (4.2), no $u \in \mathfrak{U}_*$ $(u \neq 0)$ is orthogonal to $\mathfrak{P}$, and therefore, no $u \in \mathfrak{U}_*$ is an eigenvector of the given problem. Conversely, if for any choice of $p^1, p^2, \ldots, p^\mu \in \mathfrak{P}$ the determinant of $\{(p^i, u^k)\}$ is zero, then there is a vector $u_* \in \mathfrak{U}_*$ $(u_* \neq 0)$ such that u_* is orthogonal to $\mathfrak{P}$. In this way, we have $Pu_* = 0$ and $QAu_* = Q(\lambda_* u) = \lambda_* u_*$, which proves the theorem.

Any vector $u_* \in \mathfrak{U}_*$ $(u_* \neq 0)$ that is orthogonal to $\mathfrak{P}$ is called a *superpersistent eigenvector.* As shown above, u_* has the property that it is an eigenvector of both the base problem and the given problem, and of course of all intermediate problems as well. An eigenvalue is called a *superpersistent eigenvalue* if at least one of its corresponding eigenvectors is superpersistent. In general, to a given superpersistent λ_* there correspond $j\,(1 \leq j \leq \mu)$ linearly independent superpersistent eigenvectors, say ${u_*}^1$, ${u_*}^2, \ldots, {u_*}^j$.

We now reduce our problem to the case in which every $\lambda_i^{(0)}$ is not superpersistent. In order to do this, we consider the subspace consisting of all superpersistent eigenvectors, which we call $\mathfrak{E}$. We introduce the subspace $\mathfrak{P}' = \mathfrak{P} \oplus \mathfrak{E}$ and let Q' denote the orthogonal projection operator onto $(\mathfrak{P}')^\perp$. For a moment, we replace the given problem $QAu = \lambda u$, $u = Qu$, by $Q'Au = \lambda u$, $u = Q'u$. In the latter problem, the superpersistent eigenvalues and superpersistent eigenvectors do not appear. In fact, if u_* were a superpersistent eigenvector, it would be orthogonal to $\mathfrak{P}'$, and, therefore, it would be orthogonal to $\mathfrak{P} \subset \mathfrak{P}'$. However, we already have $u_* \in \mathfrak{E} \subset \mathfrak{P}'$, which is a contradiction.

With respect to the new problem

$$Q'Au = \lambda u, \qquad Q'u = u, \tag{1}$$

every eigenvalue of the base problem admits a distinguished choice. Therefore, there is a distinguished sequence and the method described in previous sections applies. This yields lower bounds for the eigenvalues of the new problem (1). In order to obtain lower bounds for the original problem, we only have to insert in the sequence of computed lower bounds the superpersistent eigenvalues, with appropriate multiplicities given by the number of corresponding superpersistent eigenvectors. Of course, we again use the higher eigenvalues of the base problem as bounds for the higher eigenvalues of the given problem.

Summarizing, we may say that the first method, by using distinguished choices, allows us to obtain improved bounds for any number of eigenvalues. The superpersistent eigenvalues are the only ones which refuse to be improved, as they are already the exact values.

7. The General Choice

In this section, we consider the nth intermediate problem

$$Au - \lambda u = \alpha_1 p_1 + \alpha_2 p_2 + \cdots + \alpha_n p_n, \tag{1}$$

$$(u, p_i) = 0 \qquad (i = 1, 2, \ldots, n), \tag{2}$$

where no assumptions are made on $p_1, p_2, \ldots, p_n \in \mathfrak{H}$ other than linear independence. As before, the nonpersistent eigenvalues are obtained by the rules of Section 3. We again follow the procedure indicated in [W9, pp. 30–33], which is actually exactly the same as that in Section 4, except that we do not assume (4.2). Let us now suppose that λ_* is an eigenvalue of the base problem of multiplicity μ and that $u^1, u^2, \ldots, u^\mu$ are corresponding eigenvectors. We first make the simple observation that, if an eigenvalue of the base problem has a large multiplicity, namely, $n < \mu$, then there exist nontrivial combinations of the eigenfunctions which satisfy (2) so that λ_* is always persistent. In all cases, if λ_* is a persistent eigenvalue of problem (1)–(2), the Fredholm conditions

$$\left(\sum_{i=1}^{n} \alpha_i p_i, u^k\right) = 0 \qquad (k = 1, 2, \ldots, \mu) \tag{3}$$

must be satisfied (possibly only in the trivial case $\alpha_1 = \alpha_2 = \cdots \alpha_n = 0$, which is not excluded). If these conditions (3) hold, the solution of (1)–(2) is given by

$$u = R'_{\lambda*}\left(\sum_{i=1}^{n} \alpha_i p_i\right) + \sum_{h=1}^{\mu} \beta_h u^h \tag{4}$$

$$= \sum_{i=1}^{n} \alpha_i R'_{\lambda*} p_i + \sum_{h=1}^{\mu} \beta_h u^h, \tag{5}$$

where R'_{λ_*} is defined as it was in Eq. (4.11). If we now apply the orthogonality conditions (2) to u, we obtain n equations

$$\sum_{i=1}^{n} \alpha_i(R'_{\lambda_*}p_i, p_k) + \sum_{h=1}^{\mu} \beta_h(u^h, p_k) = 0 \qquad (k = 1, 2, \ldots, n). \tag{6}$$

We have altogether in (3) and (6) $n + \mu$ equations for $n + \mu$ unknowns $\alpha_1, \alpha_2, \ldots, \alpha_n, \beta_1, \beta_2, \ldots, \beta_\mu$. In order to have nontrivial solutions, the determinant of the matrix

$$\begin{matrix} n \\ \mu \end{matrix} \begin{bmatrix} \overbrace{(R'_{\lambda_*}p_i, p_k)}^{n} & \overbrace{(p_i, u^h)}^{\mu} \\ (u^h, p_k) & 0 \end{bmatrix} \qquad \begin{matrix} (i, k = 1, 2, \ldots, n) \\ (h = 1, 2, \ldots, \mu) \end{matrix} \tag{7}$$

must be zero. If this determinant is zero, then Eqs. (3) and (6) admit a certain number of independent solutions equal to the nullity of (7). Using these solutions of (3) and (6), we obtain v solutions of the form (5). Of course, the multiplicity of λ_* is given by the number of independent functions thus obtained and therefore is certainly less than or equal to v.

In buckling problems (Section 12), the first eigenvalue is the most important one. In this case, the multiplicity of the *lowest* λ_* is irrelevant since either λ_* or the lowest nonpersistent eigenvalue is the required lower bound. If v is small, the determination of the multiplicity in any case is practically feasible, but, for large v, would seemingly present difficulties. (Of course, the nonvanishing of (7) would guarantee the nonpersistency of the corresponding eigenvalue.)

Let us also observe that, by the principle of monotonocity (2.5.3), the multiplicity of λ_* in the nth intermediate problem cannot exceed $M - s$, where $M = \max\{j \,|\, \lambda_j^{(0)} = \lambda_*\}$ and s is the number of eigenvalues of the nth intermediate problem that are strictly less than λ_*. For instance, if $\lambda_1^{(0)}$ is persistent, its multiplicity can never increase.

As mentioned before, in order to determine lower bounds for the given problem, we need to find all eigenvalues, persistent and nonpersistent, preceding λ_*. The above may be helpful in determining the multiplicity of a limited number of eigenvalues.

The following result of Bazley and Fox [BF2] covers all cases.

Theorem 1. *The multiplicity of λ_* in the intermediate problem* (1)–(2) *is equal to v, the nullity of the matrix* (7).

PROOF. Let

$$a^{(1)} = [\alpha_1^{(1)}, \alpha_2^{(1)}, \ldots, \alpha_n^{(1)}, \beta_1^{(1)}, \beta_2^{(1)}, \ldots, \beta_\mu^{(1)}],$$
$$a^{(2)} = [\alpha_1^{(2)}, \alpha_2^{(2)}, \ldots, \alpha_n^{(2)}, \beta_1^{(2)}, \beta_2^{(2)}, \ldots, \beta_\mu^{(2)}],$$
$$\vdots$$
$$a^{(\nu)} = [\alpha_1^{(\nu)}, \alpha_2^{(\nu)}, \ldots, \alpha_n^{(\nu)}, \beta_1^{(\nu)}, \beta_2^{(\nu)}, \ldots, \beta_\mu^{(\nu)}]$$

be the ν linearly independent solutions of (3) and (6) and let

$$v^{(j)} = \sum_{i=1}^{n} \alpha_i^{(j)} R'_{\lambda_*} p_i + \sum_{i=1}^{\mu} \beta_i^{(j)} u^i \qquad (j = 1, 2, \ldots, \nu)$$

be the corresponding solutions of (1)–(2). Suppose there exist $\gamma^{(1)}, \gamma^{(2)}, \ldots, \gamma^{(\nu)}$, not all zero, such that

$$\sum_{j=1}^{\nu} \gamma^{(j)} v^{(j)} = 0. \tag{8}$$

Using the important fact (see Section 4) that the vectors $\sum_{i=1}^{n} \alpha_i^{(j)} R'_{\lambda_*} p_i$ are orthogonal to $\sum_{i=1}^{\mu} \beta_i^{(j)} u^i$, Eq. (8) implies two separate equations, namely

$$\sum_{j=1}^{\nu} \gamma^{(j)} \sum_{i=1}^{n} \alpha_i^{(j)} R'_{\lambda_*} p_i = 0 \tag{9}$$

and

$$\sum_{j=1}^{\nu} \gamma^{(j)} \sum_{i=1}^{\mu} \beta_i^{(j)} u^i = 0. \tag{10}$$

Since R'_{λ_*} is invertible on $\mathfrak{U}_*{}^{\perp}$, from the general principle (see Section 3), it follows that

$$\sum_{j=1}^{\nu} \gamma^{(j)} \alpha_i^{(j)} = 0 \qquad (i = 1, 2, \ldots, n). \tag{11}$$

Moreover, since the vectors u^i are independent, Eq. (10) yields

$$\sum_{j=1}^{\nu} \gamma^{(j)} \beta_i^{(j)} = 0 \qquad (i = 1, 2, \ldots, \mu). \tag{12}$$

Combining Eqs. (11) and (12), we have

$$\sum_{j=1}^{\nu} \gamma^{(j)} a^{(j)} = 0,$$

which means that $\gamma^{(1)} = \gamma^{(2)} = \cdots = \gamma^{(\nu)} = 0$, contrary to our assumption. Therefore, the $v^{(j)}$ are linearly independent.

8. Aronszajn's Rule for the Theoretical Determination of Eigenvalues

The two methods in the previous sections allow us to compute one persistent eigenvalue at a time as the zeros of certain determinants which have to be formed for each λ_* separately. Aronszajn [A1] has given a third rule which theoretically determines *all* eigenvalues, persistent and nonpersistent, using not only the zeros but also the *poles* of the Weinstein determinant (3.4). A comparison of these rules will be given in Section 9.

In order to formulate this rule, let us first observe that the determinant $W(\lambda)$, which we shall write sometimes as $W_{0n}(\lambda)$ to indicate that this determinant links the base problem to the nth intermediate problem, is a meromorphic function of λ on the lower part of the spectrum. Also, we shall denote by $W_{ij}(\lambda)$ the determinant that links the ith intermediate problem to the jth intermediate problem, $i < j$, where the ith problem plays the role of a base problem.

As usual, we define for any point λ the *order* of $W_{ij}(\lambda)$, written $\omega_{ij}(\lambda)$, as zero if $W_{ij}(\lambda)$ is finite and different from zero, as $+k$ if $W_{ij}(\lambda)$ has a zero of order k, and as $-k$ if $W_{ij}(\lambda)$ has a pole of order k ($i = 0, 1, 2, \ldots, n-1; j = 1, 2, \ldots, n$). Let us denote by $\mu_0(\lambda)$ and $\mu_n(\lambda)$ the multiplicities of λ in the spectrum of the base problem and the nth intermediate problem, respectively. (As usual, zero multiplicity means that λ is not an eigenvalue.) Then, we have the following theorem.

Aronszajn's Rule. The multiplicity $\mu_n(\lambda)$ in the nth *intermediate problem is given by*

$$\mu_n(\lambda) = \mu_0(\lambda) + \omega_{0n}(\lambda). \tag{1}$$

In other words, the multiplicity of λ remains unchanged if $W(\lambda)$ is finite and not zero, it increases at the zeros, and it decreases at the poles.

PROOF. The idea of the proof is reminiscent of what Hadamard would call the major of Huygens' principle and what is today called a semigroup property. Following Aronszajn, we show first that the rule is valid for an intermediate problem of order one. We then use the first intermediate problem as a new base problem. The use of an intermediate problem as a new base problem occurred already in the method of distinguished choices (Section 4). The second intermediate problem is now an intermediate problem of order one relative to the new base problem. Proceeding in this way, we decompose the nth intermediate problem into a chain of intermediate problems each of order one.

For the sake of simplified notations, here we only solve the first intermediate problem

$$Au - \lambda u = \alpha_1 p_1 = P_1 Au, \tag{2}$$

$$P_1 u = 0. \tag{3}$$

Let us now make some observations about the function $W_{01}(\lambda)$, written (for $p_1 = p$)

$$W_{01}(\lambda) = \sum_j \frac{(p, u_j)^2}{\lambda_j - \lambda} + \int_{\lambda_\infty - 0}^{\infty} \frac{1}{\mu - \lambda} d(E_\mu p, p). \tag{4}$$

As $(p, u_j)^2 \geq 0$ and $d(E_\mu p, p) \geq 0$ (see Definition A.28), we see that $W_{01}(\lambda)$ is positive for $\lambda < \lambda_1$ and that $dW_{01}/d\lambda$ is positive for all $\lambda < \lambda_\infty$, $W(\lambda)$ finite. Therefore, $W_{01}(\lambda)$ can only have *simple* zeros between two consecutive simple poles on the lower part of the spectrum. If λ is *not* an eigenvalue of the base problem, the determinant $W_{01}(\lambda)$ can be either zero or finite nonzero. If $W_{01}(\lambda)$ is zero, it follows from the rules of Section 3 that λ is a nonpersistent eigenvalue of (2)–(3) of multiplicity one, and, therefore, we have

$$\mu_1(\lambda) = \mu_0(\lambda) + \omega_{01}(\lambda) = 0 + 1 = 1.$$

If, however, $W_{01}(\lambda)$ is nonzero for such a λ, then λ is not an eigenvalue of (2)–(3) and we have

$$\mu_1(\lambda) = \mu_0(\lambda) + \omega_{01}(\lambda) = 0 + 0 = 0.$$

We now suppose that $\lambda = \lambda_*$ is an eigenvalue of the base problem having multiplicity $\mu_0(\lambda) \geq 1$. In this case, $W_{01}(\lambda)$ could have a zero, a simple pole, or a finite nonzero value at $\lambda = \lambda_*$. As before, let $\mathfrak{U}_* = \text{sp}\{u^1, u^2, \ldots, u^{\mu_0}\}$ denote the eigenspace of λ_*. By Fredholm's rule, Eq. (2) and (3) can have a solution only if $\alpha_1 p_1$ is orthogonal to $\mathfrak{U}_*$, in which case the general form of the solution is

$$u = \alpha_1 R'_{\lambda_*} p_1 + \sum_{i=1}^{\mu_0} \beta_i \mu^i. \tag{5}$$

Let us now consider the following three possibilities for W_{01}.

Case A. Suppose $W_{01}(\lambda_*) = \infty$. Then, $W_{01}(\lambda)$ will have a simple pole at $\lambda = \lambda_*$. Therefore, p_1 is not orthogonal to some vector in $\mathfrak{U}_*$. Thus, α_1 must be zero and the general form of the solution of (2)–(3) is given by

$$u = \sum_{i=1}^{\mu_0} \beta_i u^i, \tag{6}$$

where, because of (3), the β_i are chosen so that $(u, p_1) = 0$. This is only one condition on the β_i, so that we obtain $\mu_0 - 1$ linearly independent solutions of (2)–(3) and we have

$$\mu_1(\lambda_*) = \mu_0(\lambda_*) + \omega_{01}(\lambda_*) = \mu_0 - 1.$$

CASE B. Suppose $0 < |W_{01}(\lambda_*)| < \infty$. From the form of the resolvent (3.6), we must have here p_1 orthogonal to $\mathfrak{U}_*$. In this case, the general form of the solution is given by (5). The orthogonality condition (3) implies $\alpha_1 W(\lambda_*) = 0$. Since $W(\lambda_*) \neq 0$, α_1 is zero, and the actual solution is given by $u = \sum_{i=1}^{\mu_0} \beta_i u^i$, where the scalars β_i are completely arbitrary.

In fact, $u^1, u^2, \ldots, u^{\mu_0}$ are μ_0 independent solutions of (2)–(3). Therefore, we have

$$\mu_1(\lambda_*) = \mu_0(\lambda_*) + \omega_{01}(\lambda_*) = \mu_0 + 0 = \mu_0 .$$

CASE C. Finally, it is necessary to consider *separately* the case $W_{01}(\lambda_*) = 0$. Again, since $W_{01}(\lambda_*)$ is finite, it follows that the vector p_1 must be orthogonal to $\mathfrak{U}_*$. Therefore, $R_{\lambda_*} p_1$ is actually $R'_{\lambda_*} p_1$.

While in case B it was easy to show that α_1 in (5) is zero, we now show that, besides $u^1, u^2, \ldots, u^{\mu_0}$, there is a new eigenvector for which α_1 is *not zero*. Indeed, we have

$$AR'_{\lambda_*} p_1 - \lambda_* R'_{\lambda_*} p_1 = p_1$$

and

$$P_1 R'_{\lambda_*} p_1 = W_{01}(\lambda_*) p_1 = 0,$$

so that $R'_{\lambda_*} p_1$ satisfies both (2) and (3). It follows that $\alpha_1 R'_{\lambda_*} p_1$ is a solution for any value of α_1. Therefore, we have in

$$R'_{\lambda_*} p, u^1, u^2, \ldots, u^{\mu_0}$$

a complete set of independent solutions of (2)–(3). We can conclude that

$$\mu_1(\lambda_*) = \mu_0(\lambda_*) + \omega_{01}(\lambda_*) = \mu_0 + 1.$$

Let us note that the fact that the scalar α_1 in case C is not zero has not been sufficiently emphasized in some presentations.

We have therefore proved Aronszajn's rule for the first intermediate problem.

In order to prove the general rule (1), we proceed in like manner from the first intermediate problem to the second, from the second to the third, and so on. Our main result will follow from the decomposition

$$W_{0n}(\lambda) = W_{01}(\lambda) W_{12}(\lambda) \cdots W_{n-1,n}(\lambda). \tag{7}$$

In fact, the order of a product is equal to the sum of the orders of its factors, so that once we have proved (7) we immediately obtain

$$\begin{aligned}\mu_n(\lambda) &= \mu_{n-1}(\lambda) + \omega_{n-1,n}(\lambda)\\ &= \mu_{n-2}(\lambda) + \omega_{n-2,n-1}(\lambda) + \omega_{n-1,n}(\lambda)\\ &\ \ \vdots\\ &= \mu_0(\lambda) + \omega_{01}(\lambda) + \omega_{12}(\lambda) + \cdots + \omega_{n-1,n}(\lambda)\\ &= \mu_0(\lambda) + \omega_{0n}(\lambda).\end{aligned}$$

Indeed, we only need to show that

$$W_{qn}(\lambda) = W_{q,q+1}(\lambda)W_{q+1,n}(\lambda) \qquad (q = 0, 1, 2, \ldots, n-1),$$

from which Eq. (7) follows by induction.

We assume for the moment that

$$(p_i, p_j) = \delta_{ij} \qquad (i, j = 1, 2, \ldots, n).$$

Let us observe that if λ were not an eigenvalue of the base problem nor of any of the intermediate problems, we would have

$$W_{qn}(\lambda) = \det\begin{bmatrix}(p_{q+1}, v_{q+1}^{(q)}) & (p_{q+1}, v_{q+2}^{(q)}) & \cdots & (p_{q+1}, v_n^{(q)})\\ (p_{q+2}, v_{q+1}^{(q)}) & \cdot & \cdots & (p_{q+2}, v_n^{(q)})\\ \vdots & & & \\ (p_n, v_{q+1}^{(q)}) & \cdot & \cdots & (p_n, v_n^{(q)})\end{bmatrix}, \tag{8}$$

where $v_j^{(q)}$ is the solution of

$$\begin{gathered}Av_j^{(q)} - \sum_{i=1}^{q}(Av_j^{(q)}, p_i)p_i - \lambda v_j^{(q)} = p_j\\ (v_j^{(q)}, p_i) = 0 \qquad (i = 1, 2, \ldots, q; \quad j = q+1, q+2, \ldots, n).\end{gathered} \tag{9}$$

Similarly, we see that $v_j^{(q+1)}$ is the solution of

$$\begin{gathered}Av_j^{(q+1)} - \sum_{i=1}^{q+1}(Av_j^{(q+1)}, p_i)p_i - \lambda v_j^{(q+1)} = p_j,\\ (v_j^{(q+1)}, p_i) = 0 \qquad (i = 1, 2, \ldots, q+1; \quad j = q+2, q+3, \ldots, n).\end{gathered} \tag{10}$$

Letting $\beta_j = (Av_j^{(q+1)}, p_{q+1})$ we can write (10) as

$$Av_j^{(q+1)} - \sum_{i=1}^{q}(Av_j^{(q+1)}, p_i)p_i - \lambda v_j^{(q+1)} = p_j + \beta_j p_{q+1}$$

$$(j = q+2, q+3, \ldots, n). \tag{11}$$

Combining Eqs. (9) and (11), we obtain

$$A[v_j^{(q+1)} - v_j^{(q)}] - \sum_{i=1}^{q} (A[v_j^{(q+1)} - v_j^{(q)}], p_i)p_i - \lambda[v_j^{(q+1)} - v_j^{(q)}] = \beta_j p_{q+1}$$
$$(j = q+2, q+3, \ldots, n).$$

Therefore, we have

$$v_j^{(q+1)} - v_j^{(q)} = \beta_j v_{q+1}^{(q)} \qquad (j = q+2, q+3, \ldots, n)$$

or

$$v_j^{(q+1)} = \beta_j v_{q+1}^{(q)} + v_j^{(q)} \qquad (j = q+2, q+3, \ldots, n).$$

Note that from (10) we have $(v_j^{(q+1)}, p_{q+1}) = 0$. In (8), we now add $\bar{\beta}_j$ times the first column to the jth column $(j = 2, 3, \ldots, n-q)$ in order to obtain

$$W_{qn}(\lambda) = \det\begin{bmatrix} (p_{q+1}, v_{q+1}^{(q)}) & (p_{q+1}, v_{q+2}^{(q+1)}) & \cdots & (p_{q+1}, v_n^{(q+1)}) \\ (p_{q+2}, v_{q+1}^{(q)}) & (p_{q+2}, v_{q+2}^{(q+1)}) & \cdots & (p_{q+2}, v_n^{(q+1)}) \\ \vdots & & & \\ (p_n, v_{q+1}^{(q)}) & \cdot & \cdots & (p_n, v_n^{(q+1)}) \end{bmatrix}$$

$$= \det\begin{bmatrix} (p_{q+1}, v_{q+1}^{(q)}) & 0 & \cdots & 0 \\ (p_{q+2}, v_{q+1}^{(q)}) & (p_{q+2}, v_{q+2}^{(q+1)}) & \cdots & (p_{q+2}, v_n^{(q+1)}) \\ \vdots & & & \\ (p_n, v_{q+1}^{(q)}) & (p_n, v_{q+2}^{(q+1)}) & \cdots & (p_n, v_n^{(q+1)}) \end{bmatrix}$$

$$= (p_{q+1}, v_{q+1}^{(q)}) W_{q+1,n}(\lambda)$$

$$= W_{q,q+1}(\lambda) W_{q+1,n}(\lambda). \tag{12}$$

Since these functions are analytic, Eq. (12) holds for all $\lambda < \lambda_\infty$. Up to now, we have assumed that $(p_i, p_j) = \delta_{ij}$ $(i, j = 1, 2, \ldots, n)$. However, Aronszajn's rule holds for any basis, say $\hat{p}_1, \hat{p}_2, \ldots, \hat{p}_n$ of $\mathfrak{P}_n$. In fact, we have

$$\hat{p}_i = \sum_{j=1}^{n} \gamma_{ij} p_j,$$

so that by substituting into the determinant we obtain

$$\hat{W}_{0n}(\lambda) = [\det\{\gamma_{ij}\}]^2 W_{0n}(\lambda).$$

Since the factor is a positive constant, it follows that the poles and zeros with their respective orders are preserved.

9. Comparison of the Various Rules

We have given in Sections 4–8 three different rules for solving intermediate problems. It will be seen in Chapter 5 that all of these rules will be valid for problems of the second type as well.

For nonpersistent eigenvalues, all three methods require the determination of the nonpersistent zeros of the same determinant $W(\lambda)$. For persistent eigenvalues, there are some essential differences. We first discuss those points pertinent to numerical applications, for which intermediate problems were introduced in the first place.

The distinguished choice, which includes the special choice (see Section 5.4), led to the first numerical solution of some important problems. Whichever of the three methods of Sections 4–8 is used, one has to compute the zeros of $W(\lambda)$. In the third method, one also has to determine the *order* of $W(\lambda)$ at the eigenvalues of the base problem. Even in those cases when $W(\lambda)$ appears in closed form (expressed in terms of known functions), the determination of the *order* by computational machinery becomes inaccurate. These difficulties are due to the fact that the elements of $W(\lambda)$ are usually transcendental functions which have a meromorphic character on the lower part of the spectrum, see for instance (3.6). The determinant $W(\lambda)$ consists of sums and products of such elements, which give rise to indeterminate forms (like $0 \times \infty$). For this reason, contrary to a widely held view [12a, 19], *Aronszajn's rule* has been numerically applied *only once*, see Table B.1. Such indeterminate forms cannot occur in other methods.

Turning now to some more theoretical questions, let us establish the connection between Aronszajn's rule, which deals with the order of $W(\lambda)$, and the other rules, in which the nullity of matrices plays a part. Generally speaking, the nullity of a matrix is not the same as the order of the zero of its determinant, even for diagonal matrices. This can be seen for the simple matrix ($\lambda_1 \neq \lambda_2$)

$$\begin{bmatrix} (\lambda - \lambda_1)^2 & 0 \\ 0 & (\lambda - \lambda_2) \end{bmatrix}$$

in which for $\lambda = \lambda_1$ the nullity is one while the order is two.

Nevertheless, the consistency of the different rules is due to the fact that the matrix of $W(\lambda)$ can be replaced by a diagonal matrix with the diagonal elements $W_{01}(\lambda), W_{12}(\lambda), \ldots, W_{n-1,n}(\lambda)$, which is Aronszajn's decomposition (8.7) and plays an essential role here. Each of these elements may only have the order $+1$, 0, or -1. Using this fact, let us consider first a nonpersistent eigenvalue λ, which in the nth intermediate problem has multiplicity $\mu_n \leq n$. In this case, exactly μ_n terms on the diagonal are zero and the nullity of the matrix of $W(\lambda)$ is μ_n, in agreement with the rule of Section 7.

Now, let λ_* be an eigenvalue of the base problem having multiplicity μ_0. We consider the nth intermediate problem for $n \geq \mu_0$, using a distinguished choice, say $p^1, p^2, \ldots, p^{\mu_0}$, and any additional orthonormal vectors, say $p_{\mu_0+1}, p_{\mu_0+2}, \ldots, p_n$, which are taken to be orthogonal to $\mathfrak{U}_*$. Since $p^1, p^2, \ldots, p^{\mu_0}$ is a distinguished choice, λ_* is not an eigenvalue of the μ_0th intermediate problem and therefore the order of $W_{0\mu_0}$ is $-\mu_0$. However, from Aronszajn's decomposition (8.7), we have the respective orders given by

$$W_{0n}(\lambda_*) = \omega_{0\mu_0}(\lambda_*) + \omega_{\mu_0 n}(\lambda_*) = -\mu_0 + \omega_{\mu_0 n}(\lambda_*).$$

Therefore, by Aronszjn's rule, the multiplicity of λ_* in the nth intermediate problem is given by

$$\mu_n = \mu_0 + \omega_{0n}(\lambda_*) = \mu_0 - \mu_0 + \omega_{\mu_0 n}(\lambda_*) = \omega_{\mu_0 n}(\lambda_*).$$

Since for our choice of $\{p_{\mu_0+i}\}$ the matrix of $W_{\mu_0 n}(\lambda)$ is diagonal, the order $W_{\mu_0 n}(\lambda_*)$ is exactly the nullity of the matrix, which shows the connection between Aronszajn's rule and Weinstein's rule for the distinguished choice.

As to the general choice, the connection with Aronszajn's rule, say for the first intermediate problem, is the following. Case A of Aronszajn's rule corresponds to (7.5) with $\alpha_1 = 0$. For case B, the nullity of (7.7) is μ_0, and for case C, the nullity of (7.7) is $\mu_0 + 1$.

Finally, let us add another observation about the rules for distinguished and general choices. We assume that $n > \mu$, that $p_1, p_2, \ldots, p_\mu$ is a distinguished choice for λ_*, and that $p_{\mu+1}, p_{\mu+2}, \ldots, p_n$ are already all orthogonal to $\mathfrak{U}_*$. For this case, we can rewrite (7.7) as

$$\begin{array}{c} \mu \\ n-\mu \\ \mu \end{array}\begin{bmatrix} \overbrace{(R'_{\lambda_*} p_i, p_k)}^{\mu} & \overbrace{(R'_{\lambda_*} p_i, p_k)}^{n-\mu} & \overbrace{(p_i, u^k)}^{\mu} \\ (R'_{\lambda_*} p_i, p_k) & (R'_{\lambda_*} p_i, p_k) & 0 \\ (u^i, p_k) & 0 & 0 \end{bmatrix}. \qquad (1)$$

Developing the determinant of this matrix (1) by Laplace's rule, using the last μ columns, we see that (up to an irrelevant sign) the determinant is $\det\{(p_i, u^k)\}$ (which is in the upper right-hand corner) multiplied by the determinant of the $n \times n$ matrix in the lower left-hand corner of (1).

This last determinant in turn can be obtained by Laplace's rule, this time with respect to the last μ rows. Putting the determinant of (1) equal to zero, we obtain the equation

$$[\det\{(p_i, u^k)\}]^2 W(\lambda_*) = 0.$$

Since $\det\{(p_i, u^k)\} \neq 0$, we have precisely the results of Section 7, namely that the nullity of $W(\lambda_*)$ determines the multiplicity of the persistent value λ_* in the spectrum of the nth intermediate problem, for $\mu < n$.

10. Weinberger–Bazley–Fox Method of Truncation

In this section, we shall discuss one of the methods which arose in the numerical applications of the theory.

The method of truncation reduces the difficulties mentioned in Section 9 by constructing from the base problem a new base problem relative to which the intermediate problems are easier to solve numerically. The idea of truncation was introduced by Weinberger [W5] in problems of the first type and was developed by Bazley and Fox [BF2] for problems of the second type. All results are applicable to both types.

Let us now construct a new base problem having a finite number of distinct eigenvalues. We begin by considering the spectral representation of the original base operator A, namely

$$Au = \sum_i \lambda_i(u, u_i)u_i + \int_{\lambda_\infty - 0}^{\infty} \lambda \, dE_\lambda u.$$

For a fixed positive integer N, we define the *truncation operator of order N* by

$$T_N u = \sum_{i=1}^{N} \lambda_i(u, u_i)u_i + \lambda_{N+1} \int_{\lambda_{N+1} - 0}^{\infty} dE_\lambda u, \tag{1}$$

where we assume without loss of generality that $\lambda_N < \lambda_{N+1}$. If we let U_N be the orthogonal projection operator onto $\mathfrak{U}_N = \mathrm{sp}\{u_1, u_2, \ldots, u_N\}$, we see that

$$T_N = AU_N + \lambda_{N+1}(I - U_N). \tag{2}$$

It is clear from either (1) or (2) that the inequalities

$$(T_1 u, u) \leq (T_2 u, u) \leq \cdots \leq (T_N u, u) \leq \cdots \leq (Au, u) \tag{3}$$

hold for all $u \in \mathfrak{D}(A)$. Let us note in passing that from (2) we have

$$T_N = U_N(A - \lambda_{N+1} I)U_N + \lambda_{N+1} I. \tag{4}$$

Since $A - \lambda_{N+1} I$ is just a "shift" of A to the left, we see from (4) that each truncation operator consists of a negative-semidefinite operator of finite rank plus a multiple of the identity.

From inequalities (3), we see that the truncation operator T_N is dominated by the original base operator and the part of T_N in $\mathfrak{Q}$ is dominated by the part of A in $\mathfrak{Q}$. By the principle of monotonicity (2.5.3), we see that the eigenvalues of the intermediate operators defined with respect to T_N provide lower bounds for the eigenvalues of the part of T_N in $\mathfrak{Q}$ and therefore for those of the part of A in $\mathfrak{Q}$.

The advantage gained by using the truncation operator is that the resolvent of T_N has the form

$$R_\lambda(T_N)p = \sum_{i=1}^{N} \frac{(p, u_i)u_i}{\lambda_i - \lambda} + \frac{1}{\lambda_{N+1} - \lambda}\left[p - \sum_{i=1}^{N}(p, u_i)u_i\right].$$

Therefore, the determinant $W_n(\lambda)$ is a rational function

$$\det\left\{\sum_{i=1}^{N} \frac{(p_j, u_i)(u_i, p_k)}{\lambda_i - \lambda} + \frac{1}{\lambda_{N+1} - \lambda}\left[(p_j, p_k) - \sum_{i=1}^{N}(p_j, u_i)(u_i, p_k)\right]\right\}$$

instead of a (generally) transcendental function, thus reducing the difficulty of determining its zeros for instance, in using a general choice.

11. Convergence

In the preceding sections, we have constructed and investigated the intermediate problems $Q_1 A Q_1$, $Q_2 A Q_2, \ldots$ and have seen how the corresponding eigenvalues satisfy the inequalities

$$\lambda_k^{(0)} \le \lambda_k^{(1)} \le \cdots \le \lambda_k^{(n)} \le \cdots \le \lambda_k^{(\infty)} \qquad (k = 1, 2, \ldots), \tag{1}$$

where the $\lambda_k^{(\infty)}$ are the unknown eigenvalues of the given problem QAQ. A natural question arises as to whether or not the lower bounds $\lambda_k^{(n)}$ can be made arbitrarily close to $\lambda_k^{(\infty)}$ by increasing the order n of the intermediate problem. While we have given a partial answer in Section 6, namely in the special case of superpersistent eigenvalues, we shall now discuss this question in more detail.

From the standpoint of numerical computations, this matter is of little importance since a sequence of Rayleigh–Ritz upper bounds $\Lambda_1 \le \Lambda_2 \le \cdots$ is usually readily computable. If for large n the value $\lambda_k^{(n)}$ is not very close to Λ_k, it is little comfort to know that $\lambda_k^{(n)} \to \lambda_k^{(\infty)}$ as $n \to \infty$. On the other hand, even if there is no convergence of lower or upper bounds, the upper and lower bounds may nevertheless differ by an amount considered to be negligible for a given problem. In numerical analysis, convergence is often considered as a substitute for error estimates, but this question is of much more interest from the theoretical viewpoint since it does not give any information for numerical applications.

In any case, the problem of convergence is not trivial and has not been solved in general. However, we have the following convergence theorem of Aronszajn and Weinstein [AW1] for the case of a negative- (or positive-) definite compact operator.

Theorem 1. *Let A be a negative-definite compact operator on $\mathfrak{H}$ and let Q be an orthogonal projection operator onto a subspace $\mathfrak{Q}$ of $\mathfrak{H}$. Let $p_1, p_2, \ldots$ be a sequence complete in $\mathfrak{P} = \mathfrak{Q}^\perp$ and let Q_n be the orthogonal projection operator onto the orthogonal complement of* $\mathrm{sp}\{p_1, p_2, \ldots, p_n\}$. *Let $\lambda_1 \le \lambda_2 \le \cdots$, $\lambda_1^{(\infty)} \le \lambda_2^{(\infty)} \le \cdots$, and $\lambda_1^{(n)} \le \lambda_2^{(n)} \le \cdots$ denote the nonzero eigenvalues of A, QAQ, and Q_nAQ_n, respectively. Then for every index k ($k = 1, 2, \ldots$), we have*

$$\lim_{n\to\infty} \lambda_k^{(n)} = \lambda_k^{(\infty)}.$$

Our proof will be a slight simplification of that given in [AW1].

PROOF. Let k be given and let $\mathfrak{U}_{k-1}^{(\infty)}$ be the subspace generated by the first $k-1$ (nontrivial) eigenvectors of QAQ. We consider the variational problems

$$\hat{\lambda}_j^{(n)} = \min_{\substack{u \perp \mathfrak{P}_n \\ u \perp \mathfrak{U}_{k-1}^{(\infty)}}} R(u) \qquad (n = 1, 2, \ldots). \tag{2}$$

By Weyl's inequality (3.1.2), we have

$$\hat{\lambda}_k^{(n)} \le \lambda_k^{(n)} \qquad (n = 1, 2, \ldots). \tag{3}$$

We denote by $\hat{u}_k^{(1)}, \hat{u}_k^{(2)}, \ldots, \hat{u}_k^{(n)}, \ldots$ a sequence of minimizing vectors for (2) and we assume without loss of generality that $(\hat{u}_k^{(n)}, \hat{u}_k^{(n)}) = 1$ ($n = 1, 2, \ldots$). There exists a subsequence, say $\hat{u}_k^{(n')}$, converging weakly (see Definition A.14) to $\hat{u}_k$ in $\mathfrak{H}$. It is clear that $\hat{u}_k$ is orthogonal to $\mathfrak{U}_{k-1}^{(\infty)}$ and since $p_1, p_2, \ldots$ is complete in $\mathfrak{P}$, $\hat{u}_k$ is also orthogonal to $\mathfrak{P}$. Moreover, by (3) and the principle of monotonicity, we have

$$\hat{\lambda}_k^{(1)} \le \hat{\lambda}_k^{(2)} \le \cdots \le \lambda_k^{(\infty)} < 0.$$

Therefore, the sequence $\hat{\lambda}_k^{(n)}$ has a unique limit point $\hat{\lambda}_k^{(\infty)} < 0$. Noting that A is compact, we can now write

$$\begin{aligned} \lim_{n\to\infty} \hat{\lambda}_k^{(n)} &= \lim_{n'\to\infty} \hat{\lambda}_k^{(n')} \\ &= \lim_{n'\to\infty} (A\hat{u}_k^{(n')}, \hat{u}_k^{(n')}) \\ &= (A\hat{u}_k, \hat{u}_k) = \hat{\lambda}_k^{(\infty)} \le \lambda_k^{(\infty)} < 0, \end{aligned}$$

from which it follows that $\hat{u}_k \neq 0$. Moreover, from a general property of weakly convergent sequences, we have

$$(u_k, u_k) \leq \liminf (u_k^{(n')}, u_k^{(n')}) = 1,$$

which means (since A is negative) that

$$R(\hat{u}_k) \leq (A\hat{u}_k, \hat{u}_k).$$

Therefore, in view of inequality (3), we can conclude that

$$\begin{aligned} \lambda_k^{(\infty)} &= \min_{\substack{u \perp \mathfrak{P} \\ u \perp \mathfrak{A}_{k-1}^{(\infty)}}} R(u) \leq R(\hat{u}_k) \leq (A\hat{u}_k, \hat{u}_k) \\ &= \hat{\lambda}_k^{(\infty)} = \lim_{n\to\infty} \hat{\lambda}_k^{(n)} \\ &\leq \lim_{n\to\infty} \lambda_k^{(n)} \leq \lambda_k^{(\infty)}, \end{aligned}$$

so that equality must hold throughout and, in particular, $\lim_{n\to\infty} \lambda_k^{(n)} = \lambda_k^{(\infty)}$.

This theorem also holds in the case when A^{-1} is a positive (or negative) compact operator. The proof has to be modified in the following manner.

The existence of the minimizing vectors in (2) follows from Theorem 3.3.1. We normalize such vectors by putting $(A\hat{u}_k^{(n)}, A\hat{u}_k^{(n)}) = 1$ $(n = 1, 2, \ldots)$. The sequence $\{A\hat{u}_k^{(n)}\}$ has a subsequence, say $\{A\hat{u}_k^{(n')}\}$, converging weakly to $\hat{v}_k$ in $\mathfrak{H}$. Since A^{-1} is a positive compact operator, we have $A^{-1}A\hat{u}_k^{(n')}$ converging (strongly) to $A^{-1}\hat{v}_k = \hat{u}_k$, that is $\hat{u}_k^{(n')} \to \hat{u}_k$, so that $\hat{u}_k$ is orthogonal to $\mathfrak{P}$ and to $\mathfrak{U}_{k-1}^{(\infty)}$.

Letting $v_k^{(n')} = Au_k^{(n')}$, we can write

$$\begin{aligned} \lim_{n\to\infty} \hat{\lambda}_k^{(n)} &= \lim_{n'\to\infty} \hat{\lambda}_k^{(n')} \\ &= \lim_{n'\to\infty} \frac{(Au_k^{(n')}, u_k^{(n')})}{(u_k^{(n')}, u_k^{(n')})} \\ &= \lim_{n'\to\infty} \frac{(v_k^{(n')}, A^{-1}v_k^{(n')})}{(A^{-1}v_k^{(n')}, A^{-1}v_k^{(n')})} \\ &= \frac{(\hat{v}_k, A^{-1}\hat{v}_k)}{(A^{-1}\hat{v}_k, A^{-1}\hat{v}_k)} \\ &= \frac{(A\hat{u}_k, \hat{u}_k)}{(\hat{u}_k, \hat{u}_k)} = \hat{\lambda}_k^{(\infty)} \leq \lambda_k^{(\infty)}. \end{aligned}$$

But we also have

$$\lambda_k^{(\infty)} = \min_{\substack{u \perp \mathfrak{P} \\ u \perp \mathfrak{A}_{k-1}^{(\infty)}}} R(u) \leq R(\hat{u}_k) = \hat{\lambda}_k^{(\infty)} \leq \lambda_k^{(\infty)}$$

so that

$$\lim_{n \to \infty} \lambda_k^{(n)} = \lambda_k^{(\infty)},$$

as before.

The proof can also be modified to treat the case when A is bounded below and A^{-1} is compact.

The question of convergence of the eigenvalues of the intermediate problems which use a truncated base operator can also be answered in the following way, provided the *original* base operator is compact and negative (or positive). Let us note in passing that a truncated operator is almost always noncompact and can only be compact in the somewhat trivial case in which the original base operator has only a *finite* number of negative (or positive) eigenvalues in the lower (upper) part of its spectrum and the order of truncation exceeds the number of available negative (positive) eigenvalues.

Theorem 2. *Let A be a negative compact operator on $\mathfrak{H}$, let T_N be a truncation of A of order N, and let $\lambda_1^{(n,N)} \leq \lambda_2^{(n,N)} \leq \cdots$ denote the (nontrivial) eigenvalues of the intermediate problems*

$$Q_n T_N Q_n u = \lambda u.$$

Then, for every index k ($k = 1, 2, \ldots$), we have

$$\lim_{N \to \infty} \lambda_k^{(n,N)} = \lambda_k^{(n)}.$$

Proof. We begin by noting that the operator $A - T_N$ has the representation

$$[A - T_N]u = \sum_{i=N+1}^{\infty} (\lambda_i - \lambda_{N+1})(u, u_i)u_i.$$

Since $\lambda_{N+1} \leq \lambda_i \leq 0$ ($i = N+1, N+2, \ldots$), we see that

$$0 \leq ([A - T_N]u, u)/(u, u) \leq -\lambda_{N+1}.$$

Therefore, we have the Rayleigh quotient inequalities

$$\frac{(T_N u, u)}{(u, u)} \leq \frac{(Au, u)}{(u, u)} \leq \frac{(T_N u, u)}{(u, u)} - \lambda_{N+1}.$$

By the principle of monotonicity, the corresponding intermediate eigenvalues must satisfy

$$\lambda_k^{(n,N)} \le \lambda_k^{(n)} \le \lambda_k^{(n,N)} - \lambda_{N+1} \qquad (k = 1, 2, \ldots).$$

But $\lambda_{N+1} \to 0$ as $N \to \infty$, which proves the theorem.

If A is negative-definite, the above theorem could be combined with Theorem 1 to yield convergence to the unknown values, namely

$$\lim_{n\to\infty} \lim_{N\to\infty} \lambda_k^{(n,N)} = \lambda_k^{(\infty)} \qquad (k = 1, 2, \ldots).$$

12. Computation of Lower Bounds for the Buckling of a Clamped Plate

In 1935–1937 Weinstein gave lower bounds for the buckling load and vibrations of the clamped square plate by introducing intermediate problems in differential form. In solving these problems he used the classical technique of Lagrange multipliers and natural boundary conditions. While such techniques are still used (see Section 6.2), in the following we shall also give simultaneously a theoretical foundation by using compact operators.

Given a plane domain Ω having boundary $\partial\Omega$, we consider the eigenvalue problem (buckling problem)

$$\Delta\,\Delta w + \mu\,\Delta w = 0 \tag{1}$$

subject to the boundary conditions

$$w = \partial w/\partial n = 0, \tag{2}$$

where $\partial/\partial n$ denotes the outward normal derivative. The differential problem (1) at first glance does not appear to be of the form $Lw = \mu w$ but rather of the form $Lw = \mu M w$. While the general theory can be easily extended to the latter case for positive M, by a change of variable we shall reduce (1) to a familiar form. Proceeding formally for the moment, we note that the Rayleigh quotient is given by

$$(\Delta w, \Delta w)/D(w, w), \tag{3}$$

where (w, v) again denotes double integration over Ω and $D(w, v)$ denotes the *Dirichlet integral*, namely

$$D(w, v) = \iint_\Omega (w_x v_x + w_y v_y)\, dx\, dy.$$

Under the condition that v vanishes on $\partial\Omega$, we have Green's formula

$$D(w, v) = -(\Delta w, v).$$

Formally, the boundary condition $\partial w/\partial n = 0$ can be replaced by the condition $\oint_{\partial\Omega} p(\partial w/\partial n)\, ds = 0$ for all functions $p(s)$.

If we now impose *less stringent* conditions, namely $w = 0$ on $\partial\Omega$ and $\oint_{\partial\Omega} p_k(\partial w/\partial n)\, ds = 0$ $(k = 1, 2, \ldots, n)$, with arbitrary $p_k(s)$, the minimum of (3) under the new conditions cannot be greater than the minimum of (3) under the original boundary conditions (2). We obtain the Euler equation and boundary conditions for the new minimum problem by putting the first variation of

$$\iint_\Omega (\Delta w)^2\, dx\, dy - \mu D(w, w) - 2 \sum_{k=1}^{n} \alpha_k \oint_{\delta\Omega} p_k(\partial w/\partial n)\, ds$$

equal to zero, where $\mu, \alpha_1, \alpha_2, \ldots, \alpha_n$ are Lagrangian multipliers. This leads to the equation

$$\iint_\Omega \delta w(\Delta\, \Delta w + \mu\, \Delta w)\, dx\, dy + \oint_{\delta\Omega} \left(\Delta w - \sum_{k=1}^{n} \alpha_k p_k\right) \frac{d\,\delta w}{dn}\, ds$$

$$- \oint_{\delta\Omega} \delta w \left(\frac{d\,\Delta w}{dn} + \mu \frac{dw}{dn}\right) ds = 0.$$

Since the variations δw and $d\,\delta w/dn$ are arbitrary functions with $\delta w = 0$ on $\partial\Omega$, we see that the *Euler equation* is

$$\Delta\, \Delta w + \mu\, \Delta w = 0 \qquad \text{in} \quad \Omega$$

subject to the boundary condition $w = 0$ and the *natural boundary condition* $\Delta w = \sum_{k=1}^{n} \alpha_k p_k$ on $\partial\Omega$ and the conditions $\oint_{\partial\Omega} p_k(\partial w/\partial n)\, ds = 0$ $(k = 1, 2, \ldots, n)$. This Euler equation shows that $\Delta w + \mu w$ is a *harmonic function* in Ω. Since w vanishes on $\partial\Omega$, we must have $\Delta w + \mu w = \sum_{k=1}^{n} \alpha_k p_k$ on $\partial\Omega$. These boundary values *uniquely* determine $\Delta w + \mu w$ so we obtain the differential problem

$$\Delta w + \mu w = \sum_{k=1}^{n} \alpha_k p_k(x, y) \qquad \text{in} \quad \Omega \tag{4}$$

where $w = 0$ on $\partial\Omega$ and $\oint_{\partial\Omega} p_k(\partial w/\partial n)\, ds = 0$ $(k = 1, 2, \ldots, n)$. Here the $p_k(x, y)$ are the harmonic extensions of the $p_k(s)$.

Putting all $p_k = 0$ we have the still less stringent problem (*base problem*) $\Delta w + \mu w = 0$ in Ω, $w = 0$ on $\partial\Omega$, which is the problem of the vibrating *membrane*.

Before solving the *intermediate problem* (4) for the square $-\pi/2 \le x$ $y \le \pi/2$, let us now outline the theoretical foundations of the procedure.

Let p be a sufficiently regular harmonic (potential) function on Ω. From Green's identity we have

$$\iint\limits_{\Omega} (p\,\Delta w - w\,\Delta p) = \oint_{\delta\Omega} \left(p\,\frac{\partial w}{\partial n} - w\,\frac{\partial p}{\partial n}\right) ds$$

from which it follows that the conditions (2) can be replaced by the new conditions $w = 0$ in Ω and $(p, \Delta w) = 0$ for all harmonic functions p.

We now introduce Green's operator G which is the inverse of Δ (see Theorems A.11–A.12). Letting $u = \Delta w$ and $w = Gu$, we see that the minimum of (3) subject to the condition $(p, \Delta w) = 0$ leads to the variational problem for the Rayleigh quotient $R(u) = (Gu, u)/(u, u)$,

$$\min_{u \perp \mathfrak{P}} R(u),$$

where $\mathfrak{P}$ is the subspace of harmonic functions in $\mathscr{L}_2(\Omega)$, i.e. $p \in \mathfrak{P}$ means that p is regular in Ω and at worst has singularities on $\partial\Omega$. Of course, this problem is more general than (1) and (2) since it can be formulated for any class of finitely connected domains having no isolated boundary points and functions $u \in \mathscr{L}_2(\Omega)$. However, it has been shown that these two problems are equivalent for a large class of domains (see [AW2, p. 640].)

The corresponding *intermediate problems* yielding lower bounds are obtained by selecting a sequence $\{p_i\} \subset \mathfrak{P}$ and considering the variational problem

$$\min_{u \perp p_1, p_2, \ldots, p_n} R(u).$$

Therefore, the corresponding nth intermediate problem will have the eigenvalue equation [see Eq. (2.4)]

$$Gu - P_n Gu = \lambda u, \qquad P_n u = 0. \tag{5}$$

Let us now show that (5) is equivalent to the differential equation (4). It should be emphasized here that it is not possible to do the numerical computations using G since the Green's function for a square is extremely complicated. In the following let λ be an eigenvalue of (5) having eigenfunction u. Since G is negative definite (Theorem A.12) we know that zero cannot be an eigenvalue of (5).

From the elementary properties of the logarithmic potential, (see [18]) and Schwarz's inequality, the following facts are easy to verify. Since

$u \in \mathscr{L}_2(\Omega)$, it follows that Gu is continuous in $\Omega \cup \partial\Omega$ and therefore, from the eigenvalue equation (5), u must be continuous in $\Omega \cup \partial\Omega$. Moreover, $Gu = 0$ on $\partial\Omega$, so that from (5) we must have on $\partial\Omega$

$$u = -\lambda^{-1} P_n Gu = \sum_{i=1}^{n} \alpha_i p_i. \tag{6}$$

Equation (6) is, of course, valid pointwise on $\partial\Omega$ since both sides are continuous in $\Omega \cup \partial\Omega$. Since u is continuous and bounded on Ω, Gu is continuously differentiable on Ω, and therefore, by (5), u is continuously differentiable on Ω. Since u is continuously differentiable in Ω, Gu is twice continuously differentiable in Ω and therefore u is twice continuously differentiable in Ω. Moreover, Gu satisfies $\Delta Gu = u$ in Ω. If we now apply Δ to both sides of (5), we obtain the equation

$$-\lambda\, \Delta u + u = 0 \qquad \text{in} \quad \Omega, \tag{7}$$

subject to the orthogonality conditions

$$(u, p_i) = 0 \qquad (i = 1, 2, \ldots, n). \tag{8}$$

Putting $\mu = -\lambda^{-1}$, we rewrite (7) as

$$\Delta u + \mu u = 0 \qquad \text{in} \quad \Omega. \tag{9}$$

We now let

$$v = u - \sum_{i=1}^{n} \alpha_i p_i$$

in $\Omega \cup \partial\Omega$. Then, since $\Delta u = \Delta v$, the equation becomes

$$\Delta v + \mu v = \sum_{i=1}^{n} \mu \alpha_i p_i \qquad \text{in} \quad \Omega, \tag{10}$$

where now, in view of (6),

$$v = 0 \qquad \text{on} \quad \partial\Omega. \tag{11}$$

Moreover, the orthogonality conditions (8) are transformed into

$$(\Delta v, p_i) = 0 \qquad (i = 1, 2, \ldots, n) \tag{12}$$

since we have

$$(u, p_i) = \lambda(\Delta u, p_i) = \lambda(\Delta v, p_i) \qquad (i = 1, 2, \ldots, n).$$

For $p_i = 0$ $(i = 1, 2, \ldots, n)$, we have the old *base problem*

$$\Delta v + \mu v = 0 \qquad \text{in} \quad \Omega,$$
$$v = 0 \qquad \text{on} \quad \partial\Omega,$$

which is again the *membrane problem*. It has known eigenvalues $j^2 + k^2$ and corresponding normalized eigenvectors

$$(2/\pi) \sin j(x + \tfrac{1}{2}\pi) \sin k(y + \tfrac{1}{2}\pi) \qquad (j, k = 1, 2, \ldots). \tag{13}$$

Let us order the eigenvalues $\{j^2 + k^2\}$ in a nondecreasing sequence $2, 5, 5, 8, \ldots$ and call the corresponding eigenfunctions $u_1, u_2, \ldots$. In order to obtain the eigenvalues of the intermediate problem (10)–(12), we introduce a sequence of harmonic functions in the following way. We take the normal derivative of each u_i $(i = 1, 2, \ldots)$ on the boundary and select the harmonic functions having these boundary values. We then eliminate all functions that depend linearly on the preceding ones and call those that remain $p_1(x, y), p_2(x, y), \ldots$. A similar construction is possible for a wide class of domains and leads to a distinguished sequence of harmonic functions (see Weinstein [W9]). The completeness (actually overcompleteness) of such sequences was proved by Colautti [C1].

Let us now consider for the sake of simplicity the first intermediate problem

$$\Delta v + \mu v = \mu \alpha_1 p_1, \qquad v = 0 \quad \text{on} \quad \partial\Omega, \tag{14}$$

where

$$p_1(x, y) = \cosh x \cos y + \cos x \cosh y. \tag{15}$$

Since p_i is a distinguished choice (Section 4) for $\mu_1^{(0)} = 2$, we know that the first eigenvalue of the first intermediate problem must be greater than 2. Therefore, in our computation, we will assume that $\mu > 2$. For μ not an eigenvalue of the base problem, the intermediate problem (14) admits the solution

$$\begin{aligned} v = &-\frac{\alpha_1 \cosh(\pi/2)}{\cos[(\pi/2)(\mu-1)^{1/2}]}\{\cos[(\mu-1)^{1/2}x]\cos y + \cos x \cos[(\mu-1)^{1/2}y]\} \\ &+ \alpha_1[\cosh y \cos x + \cosh x \cos y] \\ = {}&\alpha_1 v_1 + \alpha_1 p_1, \qquad \alpha_1 \neq 0. \end{aligned} \tag{16}$$

Using the orthogonality conditions (12) and noting that

$$\begin{aligned} (\Delta v, p_1) &= (\Delta v_1, p_1) \\ &= -\mu(v_1, p_1) \end{aligned}$$

we obtain the explicit formula for the Weinstein determinant

$$W(\mu) = \frac{\pi}{2}\left[\tanh\frac{\pi}{2}\right] + \frac{\pi(\mu-1)^{1/2}}{2}\left[\tan\frac{\pi(\mu-1)^{1/2}}{2}\right] + \frac{\mu}{2-\mu}, \tag{17}$$

where an inessential constant factor has been omitted for simplicity. It is important to note that this $W(\mu)$ appears in *closed form*.

It is easy to see that $W(\mu)$ is negative in the interval $2 < \mu \leq 5$. The smallest zero of $W(\mu)$ is $\mu = 5.1550$. However, $\mu_1^{(1)}$ is equal to 5.0 and not 5.1550 because $\mu_2^{(0)} = 5$ is a persistent eigenvalue, a fact which may be obtained by any of the three rules previously given in Sections 6–8. In fact, we have $\mu_1^{(1)} = \mu_2^{(1)} = 5.0$ and $\mu_1^{(3)} = 5.1550$, as already given in [W6]. In order to get an improved lower bound for $\mu_1^{(\infty)}$, we use a distinguished choice corresponding to $\mu_2 = 5$, namely

$$p^1(x, y) = \frac{\sin 2x \cosh 2y}{\cosh \pi} + \frac{2 \sinh x \cos y}{\sinh(\pi/2)}, \qquad p^2(x, y) = p^1(y, x).$$

Following the procedure of Section 4, we put

$$p^3(x, y) = p_1(x, y).$$

We now note that p^3 is orthogonal to $u_2^{(0)}$ and $u_3^{(0)}$, so that $R_5 = R_5{}'$. Therefore, the determinant (4.13) evaluated at 5 is just (17) at 5. But (17) at 5 is not zero so that 5 is not an eigenvalue of the *third* intermediate problem generated by p^1, p^2, p^3.

We know by monotonicity that

$$5 = \mu_1^{(1)} \leq \mu_1^{(3)}.$$

Since we also know that 5 is not an eigenvalue of the third intermediate problem, we can conclude that

$$5 < \mu_1^{(3)}.$$

Using test functions $\cos^2 x \cos^2 y$ and $\cos^3 x \cos^3 y$ Weinstein found that 5.31173 is an upper bound (see Section 2.3) for the buckling load $\mu_1^{(\infty)}$.

Therefore, we only have to find the smallest nonpersistent root of the W-determinant generated by p^1, p^2, p^3 in the interval $5 < \mu < 5.32$.

Similar to the derivation of v_1 in Eq. (16), we obtain

$$v^1(x, y) = \frac{\sin 2x \cos[y(\mu - 4)^{1/2}]}{\cos[(\pi/2)(\mu - 4)^{1/2}]} + \frac{2 \cos y \sin[x(\mu - 1)^{1/2}]}{\sin[(\pi/2)(\mu - 1)^{1/2}]},$$

$$v^2(x, y) = v^1(y, x),$$

and

$$v^3(x, y) = v_1(x, y).$$

Then, we have to consider the determinant

$$W(\mu) = \det\{(\Delta v^i, p^j)\} \qquad (i, j = 1, 2, 3),$$

which turns out to be

$$(v^1, p^1) \cdot (v^2, p^2) \cdot (v^3, p^3) = (v^1, p^1)^2 \cdot (v^3, p^3),$$

where again inessential factors are omitted. To determine the smallest nonpersistent root, we note that we have already seen that 5.1550 is the smallest root of (v^3, p^3). We have

$$(p^1, v^1) = (\pi/\mu)\{2(\tanh \pi) + (\mu - 4)^{1/2} \tan[(\pi/2)(\mu - 4)^{1/2}] \\ - [16\mu/5\pi(\mu - 5)] + 4[\coth(\pi/2)] - (\mu - 1)^{1/2} \cot[(\pi/2)(\mu - 1)^{1/2}]\},$$

which tends to $-\infty$ as $\mu \downarrow 5$, increases with μ by the general rule, is still negative at $\mu = 5.2$, and tends to $+\infty$ as $\mu \uparrow 8$. Therefore, there is a single nonpersistent root of (v^1, p^1) in the interval $5.2 < \mu < 8$, which is $\mu_2^{(3)}$. For the third intermediate problem, we have

$$\mu_1^{(3)} < \mu_2^{(3)} = \mu_3^{(3)} < 8,$$

so that indeed

$$\mu_1^{(3)} = 5.1550$$

is a lower bound for $\mu_1^{(\infty)}$.

Using higher order intermediate problems Weinstein gave the improved lower bound $5.30362 \leq \mu_1^{(\infty)}$, see also Section 13.

Later, Trefftz [T8, T9] reconsidered the same problem and referred to Weinstein's paper. However, his alleged solution contains several *essential mistakes*.† Instead of deriving the Euler equation (10), he considers only the variational problem defining an intermediate problem. Trefftz attempts to determine the corresponding minimum by introducing the function

$$\sum_{j,k=1}^{\infty} a_{jk}[\sin j(x + \tfrac{1}{2}\pi)] \sin k(y + \tfrac{1}{2}\pi)$$

into the variational problem and minimizing with respect to the unknown coefficients a_{jk}. He tacitly assumes that neither $\mu = 2$ nor $\mu = 5$ is the unknown minimum. This exclusion leads to formulas containing $\mu - 2$ or $\mu - 5$ as factors in the denominators. Obviously, Trefftz determines only *upper bounds to the lower bounds* given by a correct solution of the intermediate problems. The obtained numerical values are sometimes *above* the value for which they were supposed to give lower bounds [W27]. Unfortunately, several engineers have used, extended, and included in

† Trefftz's lack of understanding of [W6] was probably partly due to the effects of what was to be a fatal illness.

textbooks Trefftz's procedure under the name "Lagrangian multiplier method," thus obtaining numerical results open to discussion [31, BH1].

Computations by the method of intermediate problems similar to those given earlier in this section also lead to lower bounds for the frequencies of the vibration of a clamped square plate, given by the equation $\Delta\,\Delta w = \lambda w$. Weinstein [W9] obtained the inequalities

$$13.294 \leq \lambda_1 \leq 13.37.$$

Moreover, for the sake of orientation and to test the method, using only the first intermediate problem, Weinstein [W9] obtained at the same time

$$50.41 \leq \lambda_2 = \lambda_3 \leq 55.76,$$
$$112.36 \leq \lambda_4 \leq 134.56.$$

Later in 1950, by using higher intermediate problems and better computational machinery, Aronszajn and his collaborators computed lower bounds for the first 13 eigenvalues, which gave further evidence of the high precision results obtainable by the method.

Another application of the first type of intermediate problems, namely a problem connected with the Navier–Stokes equation, will be discussed in Section 7.7.

13. On the Symmetries of the Eigenfunctions

Let us return to the buckling problem. An approach to getting rid of the persistent eigenvalue 5 that is somewhat different and simpler than that of Section 12 is to take advantage of the symmetries of the solution of the given problem in the following manner [W8]. Any solution $w_1(x, y)$ of the variational problem

$$\min[(\Delta w, \Delta w)/D(w, w)] \tag{1}$$

under the conditions

$$w = \partial w/\partial n = 0 \tag{2}$$

is an even function of x (and also of y). This has the effect of raising the eigenvalues of the corresponding base problem. In fact, any solution w_1 can be written as

$$w_1(x, y) = \tfrac{1}{2}[w_1(x, y) + w_1(-x, y)] + \tfrac{1}{2}[w_1(x, y) - w_1(-x, y)]. \tag{3}$$

In view of Eq. (12.1), both the even and odd terms on the right-hand side of (3) would be minimizing functions. We shall prove that the odd part is zero, so that w_1 is actually even.

Suppose an odd function w_0 minimizes (1) over Ω under (2), and therefore that its Rayleigh quotient must be less than 5.32. Consider the rectangle $\Omega^+ = \{(x, y) | 0 < x < \pi/2, -\pi/2 < y < \pi/2\}$. Then, the Rayleigh quotient of w_0 over Ω^+ is also less than 5.32. Therefore, the minimum of (1) over Ω^+ under the conditions

$$w = 0 \quad \text{on} \quad x = 0$$

and

$$w = \partial w/\partial n = 0 \quad \text{on the other three sides}$$

cannot exceed 5.32. The relaxed variational problem on Ω^+ under the new conditions

$$w = 0, \quad x = 0, \pi/2 \tag{4}$$

and

$$\int_0^{\pi/2} (\partial w/\partial n) \sin 2x \, dx = 0, \qquad y = \pm\pi/2 \tag{5}$$

still cannot have a minimum greater than 5.32. We shall show, however, that a minimizing function $z(x, y)$ yields a Rayleigh quotient greater than 5.77, which proves that the original problem has only even solutions.

The last condition (5) is nothing other than

$$(\Delta w, p) = 0,$$

where

$$p = (\cosh 2y \sin 2x)/\cosh \pi.$$

The base problem for the latter problem is the membrane problem for Ω^+. Its first eigenvalue is 5, is simple, and has the corresponding eigenfunction

$$\sin 2x \cos y.$$

Obviously, p is a distinguished choice for 5, and therefore 5 is not persistent. We have to consider the function

$$z(x, y) = \alpha_1 \left\{ \cos[y(\mu - 4)^{1/2}] - \frac{\cos[(\pi/2)(\mu - 4)^{1/2}]}{\cosh \pi} \cos 2y \right\} \sin 2x, \tag{6}$$

which corresponds to (12.16). Therefore [see (12.17)], the minimum is given by the smallest root of

$$-\tfrac{1}{2}\pi(\mu - 4)^{1/2} \tan [\tfrac{1}{2}\pi(\mu - 4)^{1/2}] = \pi \tanh \pi.$$

This root turns out to be greater than 5.77, contradicting the assumption that it is less than 5.32. Thus, $w_0(x, y)$ must be identically zero and the minimizing function is indeed even.

Let us remark that the solution (6) of one variational problem under (4) and (5) also minimizes the same variational problem under the conditions

$$w = 0, \qquad x = 0, \pi/2, \tag{7}$$

$$w = \partial w/\partial n = 0, \qquad y = \pm\pi/2. \tag{8}$$

In fact, problem (1), (7)–(8) cannot have a smaller minimum than problem (1), (4)–(5). On the other hand, if we require the function $z(x, y)$ to satisfy (8), we obtain exactly Eq. (5) of the intermediate problem and therefore the same minimum.

This shows that the chain of intermediate problems in some cases turns out to be finite for partial (also ordinary) differential equations.

Using the evenness of the minimizing function, we can also get the improved bound of Section 12 by introducing the sequence of even harmonic functions

$$p_n(x, y) = \cosh nx \cos ny \qquad (n = 1, 2, \ldots),$$

which simplifies the computation. In this way, combining the lower bounds of the fourth intermediate problem with a Rayleigh–Ritz value, we obtain [W8]

$$5.30362 \leq \mu_1^{(\infty)} \leq 5.31173.$$

Of course, to obtain higher eigenvalues, we have to consider other symmetry classes. Similar considerations were later used in various related problems (see, for instance, Stadter [S2]).

14. Weyl's Second Lemma

About thirty years after his *first fundamental lemma* (see Section 3.1), Weyl gave in 1940 the following lemma [W33], which has become even more widely known and is usually just called *Weyl's lemma*.

A year later a second proof of this result, which we give here, was given by Weinstein [W11], who used ideas developed in the first type of intermediate problems, thereby establishing a connection between two seemingly unrelated subjects. Since that time, many other proofs of the same lemma have been given (see, for instance, [G1]).

We consider a circular disk Ω having boundary Γ and define a class of functions $\mathscr{K}_2$ by

$$\mathscr{K}_2 = \{w \in \mathscr{C}^2(\Omega \cup \Gamma) : w, w_x, w_y, w_{xy}, w_{xx}, w_{yy} = 0 \quad \text{on} \quad \Gamma\},$$

where $\mathscr{C}^2$, as usual, denotes continuous differentiability up to (and including) the second order. Our Hilbert space here is $\mathfrak{H} = \mathscr{L}_2$. Let $\mathfrak{P}$ be the subspace of $\mathscr{L}_2$ functions equivalent to functions which are regular and harmonic in Ω. Let $\mathfrak{Q} = \mathfrak{P}^\perp$. This yields the decomposition $\mathfrak{H} = \mathfrak{P} \oplus \mathfrak{Q}$ (see [W11]).

(*Weyl's Second Lemma*). *A function $u \in \mathfrak{H}$ that satisfies*

$$(u, \Delta v) = 0$$

for all $v \in \mathscr{K}_2$ is a harmonic function.

The assertion concerning u follows trivially from Green's identity (2) if it is assumed that u is equivalent to a function, say $\hat{u}$, which is *regular* in $\Omega \cup \Gamma$.

Proof. The eigenvalue problem for a clamped plate [W9]

$$\begin{aligned} \Delta\, \Delta w &= \lambda w \qquad \text{in} \quad \Omega, \\ w &= \partial w / \partial n = 0 \qquad \text{on} \quad \Gamma \end{aligned} \tag{1}$$

is explicitly solvable for a circular disk, (see [3, pp. 307–308]). The eigenfunctions $w_1, w_2, \ldots$ of (1) form a complete orthonormal system in $\mathfrak{H}$. Now, let $\hat{p}$ be any harmonic function which is regular in $\Omega \cup \Gamma$. We have as in Section 12

$$(\hat{p}, \Delta w_j) = (\Delta \hat{p}, w_j) + \oint_\Gamma \hat{p}(\partial w_j/\partial n)\, ds - \oint_\Gamma (\partial \hat{p}/\partial n) w_j\, ds = 0. \tag{2}$$

However, by a classical result of Zaremba [Z1], every $p \in \mathfrak{P}$ is the (strong) limit of functions $\hat{p}$ (regular in $\Omega \cup \Gamma$), so that we have

$$(p, \Delta w_j) = 0 \qquad (j = 1, 2, \ldots) \tag{3}$$

for all $p \in \mathfrak{P}$. In other words, each $\Delta w_j \in \mathfrak{Q}$. It follows then that the sequence $w_1, w_2, \ldots$ belonging to (1) is also a complete orthonormal system of eigenfunctions (all having nonzero eigenvalues $\lambda_1, \lambda_2, \ldots$) for the new problem (used previously in the vibration problem)

$$\begin{aligned} \Delta\, \Delta w &= \lambda w \qquad \text{in} \quad \Omega, \\ w &= 0 \qquad \text{on} \quad \Gamma, \\ (p, \Delta w) &= 0 \qquad \text{for all} \quad p \in \mathfrak{P}. \end{aligned} \tag{4}$$

We now show that the functions Δw_j are complete in $\mathfrak{Q}$. Let G denote the Green's operator for the Poisson equation

$$\Delta w = f \quad \text{in} \quad \Omega,$$
$$w = 0 \quad \text{on} \quad \Gamma.$$

Putting $f_j = \Delta w_j \; (j = 1, 2, \ldots)$, we see that $Gf_j = w_j \; (j = 1, 2, \ldots)$. Moreover, since the w_j are eigenfunctions of (1), we have

$$\begin{aligned} Gf_j = w_j &= (1/\lambda_j)\, \Delta\, \Delta w_j \\ &= (1/\lambda_j)\, \Delta(\Delta w_j + p_j) \qquad (j = 1, 2, \ldots), \end{aligned} \tag{5}$$

where each p_j is a regular harmonic function chosen so that $\Delta w_j + p_j = 0$ on Γ. We apply G to both sides of (5) to get

$$GGf_j = (1/\lambda_j)(f_j + p_j) \qquad (j = 1, 2, \ldots). \tag{6}$$

Moreover, we have, from (3), $(f_j, p) = 0$ for all $p \in \mathfrak{P}$. Letting P denote the projection operator onto $\mathfrak{P}$, we see from (6) that the f_j are eigenfunctions of

$$GGf - PGGf = (1/\lambda)f$$
$$(f, p) = 0 \qquad \text{for all} \quad p \in \mathfrak{P}.$$

Since the part of $GG - PGG$ in $\mathfrak{Q}$ is compact and selfadjoint (see Section 3.2), it has a complete system of eigenfunctions in $\mathfrak{Q}$. Suppose there is one, say f_0, which is not an f_j (or linear combination of f_j's). Letting $w_0 = Gf_0$, we have

$$\begin{aligned} \Delta\, \Delta w_0 = \Delta f_0 &= \lambda_0\, \Delta(GGf_0 - PGGf_0) \\ &= \lambda_0\, w_0, \end{aligned}$$

where

$$w_0 = 0 \qquad \text{on} \quad \Gamma$$

and

$$(\Delta w_0, p) = 0 \qquad \text{for all} \quad p \in \mathfrak{P}.$$

This contradicts the completeness of the w_j. Therefore, the sequence $f_1, f_2, \ldots$ is complete in $\mathfrak{Q}$. Since each f_j can be approximated by functions Δv with $v \in \mathscr{H}_2$, we have

$$(u, f_j) = 0 \qquad (j = 1, 2, \ldots). \tag{7}$$

In view of the completeness of $\{f_i\}$, (7) means that u must be in $\mathfrak{P}$, which completes the proof.

Chapter Five

Intermediate Problems of the Second Type

1. Formulation of Problems of the Second Type

We now turn to the consideration of a second type of intermediate problems, introduced by Aronszajn, which led to numerous modifications and numerical applications, mainly through the work of Bazley, who was soon joined by Fox. Their collaboration produced many remarkable results.

Aronszajn considered operators of the type $A + B$, where we may take at once $A \in \mathscr{S}$ and B positive (therefore symmetric). The classical study of sums of operators goes back to Weyl [W30] (for instance, Weyl's theorem —see Section 3.1—that the eigenvalues of $A + B$ dominate those of A). However, Aronszajn put the consideration of such operators into the framework of intermediate problems in the following ingenious way.

Assuming that the eigenvalues of A are known (a base problem) and again *using projections on finite-dimensional spaces*, we introduce a chain of operators $A + B_n$, where B_n has finite-dimensional range and will be called an operator of finite rank or a degenerate operator. These operators are constructed in such a way that the eigenvalues of $A + B_n$ do not decrease with n and are explicitly computable from the operator A.

To this end, let us begin by noting that the Rayleigh quotient for a given operator $A + B$ is

$$R^{(\infty)}(u) = [(Au, u) + (Bu, u)]/(u, u), \tag{1}$$

where $u \in \mathfrak{D}(A + B) = \mathfrak{D}(A) \cap \mathfrak{D}(B)$ and where the superscript ∞ has the same meaning as in Section 4.2. Observing that B is positive, Aronszajn introduced a new inner product in $\mathfrak{D}(B)$ defined by

$$[u, v] = (Bu, v).$$

Taking the closure of $\mathfrak{D}(B)$ in the norm $[u, u]^{1/2}$, we obtain a new (complete) Hilbert space $\mathfrak{H}'$.

We now select a sequence of elements $p_1, p_2, \ldots \in \mathfrak{D}(B)$. Unlike the first type of problems, the p_i are completely arbitrary for bounded operators B. To explain the basic ideas, let us suppose for the moment that

$$[p_i, p_k] = \delta_{ik} \qquad (i, k = 1, 2, \ldots).$$

For any $u \in \mathfrak{D}(B)$, we have, by *Bessel's inequality*,

$$\sum_{i=1}^{\infty} |(u, Bp_i)|^2 = \sum_{i=1}^{\infty} |[u, p_i]|^2 \leq [u, u] = (Bu, u).$$

We can replace the series on the left by a finite sum for arbitrary n, so that

$$\sum_{i=1}^{n} |(u, Bp_i)|^2 \leq (Bu, u), \qquad u \in \mathfrak{D}(B).$$

In view of the Rayleigh quotient (1), we have

$$R^{(n)}(u) = [(Au, u) + \sum_{i=1}^{n} |(u, Bp_i)|^2]/(u, u)$$
$$\leq R^{(\infty)}(u), \qquad u \in \mathfrak{D}(A + B) \subset \mathfrak{D}(B). \tag{2}$$

The eigenvalue problem corresponding to $R^{(n)}(u)$ is given by

$$Au + \sum_{i=1}^{n} (u, Bp_i)Bp_i = \lambda u.\dagger \tag{3}$$

This can be verified by standard procedures (see, for instance, Section 1.3), or simply by scalar multiplication of (3) by u, which leads back to $R^{(n)}(u)$. Let us note that in (3) the sum is just $BP^n u$, where P^n is the projection, orthogonal with respect to the new norm, of u on $\mathfrak{P}_n = sp\{p_1, p_2, \ldots, p_n\}$.

In the case in which p_i is not an orthonormal sequence, $BP^n u = B_n u$ is given by the more general expression

$$B_n u = \sum_{i=1}^{n} \sum_{j=1}^{n} (u, Bp_i)\beta_{ij} Bp_j, \tag{4}$$

† On the very first page of the Foreword of the Russian translation [12a, p. 5], editor Lidskii states that the essence of the method of intermediate problems is the consideration of the operator $Au - \sum_{k=1}^{n} (u, Ap_k)p_k$ which he supposes to give lower bounds to the eigenvalues of $A + B$. (In his notation $H = A + B$, $H^{(0)} = A$, and $H^{(n)}u = Au - \sum_{k=1}^{n} (u, Ap_k)p_k$.) This would be a remarkable result since the operator B appears nowhere in the definition of the *nonsymmetric* operator $H^{(n)}$. This fundamental confusion seems to be due to the editor's unawareness of the difference between the first and second types of intermediate problems.

where $\{\beta_{ij}\}$ is the inverse matrix of $\{[p_i, p_j]\}$ $(i, j = 1, 2, \ldots, n)$, and is therefore symmetric. Instead of (3), we obtain

$$(A + B_n)u = Au + \sum_{i=1}^{n} \sum_{j=1}^{n} (u, Bp_i)\beta_{ij} Bp_j = \lambda u. \tag{5}$$

In view of the fact that the sum in (3) or (5) is essentially a projection and cannot therefore decrease as the number of dimensions increases, we have the inequalities

$$R(u) \leq R^{(1)}(u) \leq R^{(2)}(u) \leq \cdots \leq R^{(\infty)}(u). \tag{6}$$

By a classical theorem of Weyl's [W30] (see Theorem A10), we know that the essential spectrum of $A + B_n$ is the same as that of A. However, it is conceivable that all the isolated eigenvalues of A could disappear.

Let us now show that, if A has sufficiently many isolated eigenvalues at the beginning of its spectrum to make the indices meaningful, then the spectrum of each of the operators $A + B_n$ $(n = 1, 2, \ldots)$ actually begins with isolated eigenvalues of finite multiplicity. In the following, we use the same notations $\lambda_i^{(n)}$ and $u_i^{(n)}$ for the eigenvalues and eigenvectors of $A + B_n$ as we did for those of the nth intermediate problem in the first type.

The range of the degenerate operator B_n is a finite-dimensional space $\mathfrak{S}_n$, where $\dim \mathfrak{S}_n \leq n$. The lowest point in the spectrum of the semi-bounded operator $A_n = A + B_n$ is given by $\lambda_* = \inf[(A_n u, u)/(u, u)]$, $u \in \mathfrak{D}$, $u \neq 0$ (see proof of Theorem 3.3.1). Then, we have

$$\begin{aligned} \lambda_* &= \inf_{u \in \mathfrak{D}} [(A_n u, u)/(u, u)] \leq \inf_{\substack{u \in \mathfrak{D} \\ u \perp \mathfrak{S}_n}} [(A_n u, u)/(u, u)] \\ &= \inf_{\substack{u \in \mathfrak{D} \\ u \perp \mathfrak{S}_n}} [(Au, u)/(u, u)]. \end{aligned} \tag{7}$$

However, the last quantity in (7) is actually a minimum, as was shown in Section 3.3, and by Weyl's inequality (3.1.2), we have

$$\min_{\substack{u \in \mathfrak{D} \\ u \perp \mathfrak{S}_n}} [(Au, u)/(u, u)] \leq \lambda_{n+1}, \tag{8}$$

so that

$$\lambda_* \leq \lambda_{n+1}. \tag{9}$$

If λ_* were *not* an isolated eigenvalue of finite multiplicity, it would be in the essential spectrum of A_n and therefore by Weyl's theorem also in the essential spectrum of A. In this case, we would have $\lambda_\infty \leq \lambda_* \leq \lambda_{n+1} < \lambda_\infty$, which contradicts our assumption. Therefore, we can conclude that λ_* is in fact the first isolated eigenvalue of A_n, namely $\lambda_1^{(n)}$, and satisfies the inequality

$$\lambda_1^{(n)} \leq \lambda_{n+1}. \tag{10}$$

It is remarkable that (10) looks just like (7.3.4). However, the *meaning is quite different* because, despite similar notations, we have two different problems. Moreover, it should be noted that (10) is a consequence of Weyl's inequality (3.1.2), but *not* a consequence of the minimum–maximum principle.

Let us denote by $u_1^{(n)}$ an eigenvector corresponding to $\lambda_1^{(n)}$. If we proceed in the above manner and introduce one by one the additional orthogonality conditions $u \perp u_1^{(n)}, u_2^{(n)}, \ldots$ (compare Section 3.6), we obtain the inequalities, also given by Weyl [W31] in the compact case,

$$\lambda_i^{(n)} \leq \lambda_{i+n} \qquad (i = 1, 2, \ldots), \tag{11}$$

where $i + n$ cannot exceed N in the case when A has only N initial eigenvalues.

We have constructed a chain of intermediate eigenvalue problems depending on an index n, having eigenvalues

$$\lambda_1^{(0)} \leq \lambda_2^{(0)} \leq \ldots, \quad \lambda_1^{(n)} \leq \lambda_2^{(n)} \ldots, \quad \text{and} \quad \lambda_1^{(\infty)} \leq \lambda_2^{(\infty)} \leq \ldots,$$

of A, $A + B_n$, and $A + B$, respectively. In view of inequalities (6), by Weyl's *monotonicity principle* (see Section 3.1), we have the eigenvalue inequalities

$$\lambda_i^{(0)} \leq \lambda_i^{(1)} \leq \cdots \leq \lambda_i^{(n)} \leq \cdots \leq \lambda_i^{(\infty)} \qquad (i = 1, 2, \ldots).$$

2. Finite Rank Perturbations and Intermediate Problems of the Second Type

The determination of the eigenvalues of $A + B_n$ is an essential part of the second type of intermediate problems. At first glance, $A + B_n$ appears to be a perturbation of finite rank of A. As matter of fact, in view of the flexibility of the choice of B_n and the various other requirements of the theory, the problem here is not completely identical to that of a *given, fixed* finite rank perturbation of A.

Nevertheless, it should be emphasized that the determination of the spectrum of operators perturbed by operators of finite rank was an unsolved problem until several basically different solutions were obtained as an *unexpected by-product* of intermediate problems of the second type. In turn, all solutions of the second type are patterned after those of the first type of problems. Let us note that some inconclusive attempts to determine the spectrum of perturbed operators were made, not only before, but after the solutions via intermediate problems were already obtained. See for instance Krein [K7, especially pp. 605–612].

To avoid repeating ourselves in Chapter 9, which deals in detail with perturbations, we begin here with the consideration of an arbitrary, bounded, symmetric operator D_n of rank n which is not necessarily related to a given operator B.

Let us first give in the usual way the general form of such an operator D_n. We can write

$$D_n u = \sum_{j=1}^{n} \beta_j(u) q_j,$$

where the $\beta_j(u)$ are functionals of u and the q_j form a basis for $\mathfrak{R}(D_n)$. Taking the inner product of both sides with q_k $(k = 1, 2, \ldots, n)$, we have

$$(D_n u, q_k) = \sum_{j=1}^{n} \beta_j(u)(q_j, q_k),$$

and therefore

$$\beta_j(u) = \sum_{k=1}^{n} \gamma_{kj}(D_n u, q_k) \qquad (j = 1, 2, \ldots, n),$$

where $\{\gamma_{kj}\}$ is the inverse of $\{(q_j, q_k)\}$. Since D_n is symmetric, we have

$$\beta_j(u) = \sum_{k=1}^{n} \gamma_{kj}(u, D_n q_k) \qquad (j = 1, 2, \ldots, n).$$

Moreover, for each q_k we have coefficients η_{ik} such that

$$D_n q_k = \sum_{i=1}^{n} \eta_{ik} q_i,$$

and so

$$\beta_j(u) = \sum_{k=1}^{n} \sum_{i=1}^{n} \gamma_{kj} \bar{\eta}_{ik}(u, q_i).$$

Therefore, it follows that

$$D_n u = \sum_{i=1}^{n} \sum_{j=1}^{n} \alpha_{ij}(u, q_i) q_j, \tag{1}$$

where

$$\alpha_{ij} = \sum_{k=1}^{n} \gamma_{kj} \bar{\eta}_{ik}.$$

Equation (1) corresponds to (1.4) with $\alpha_{ij} = \beta_{ij}$ and $q_i = Bp_i$. Since D_n is selfadjoint, we have $\alpha_{ij} = \bar{\alpha}_{ji}$. Using classical matrix theory, by changing the basis $q_1, q_2, \ldots, q_n$ into $q_1', q_2', \ldots, q_n'$, we can write the simplified form

$$D_n u = \sum_{i=1}^{n} \alpha_i(u, q_i')q_i', \tag{2}$$

where the set $\{q_i'\}$ is orthonormal while the α_i, as eigenvalues of a symmetric matrix, are real. Henceforth, we write q_i instead of q_i' in Eq. (2), $(q_i, q_j) = \delta_{ij}$.

We now proceed to solve the eigenvalue problem

$$A_n u = Au + D_n u = \lambda u, \tag{3}$$

which we write in the form

$$Au - \lambda u = -\sum_{i=1}^{n} \alpha_i(u, q_i)q_i. \tag{4}$$

Our procedure will parallel that of Chapter 4.

If λ is not in the spectrum of A, that is, if λ is a *nonpersistent eigenvalue*, we obtain

$$u = -\sum_{i=1}^{n} \alpha_i(u, q_i)R_\lambda q_i. \tag{5}$$

Taking the inner product of both sides of (5) with q_j, we have

$$(u, q_j) = -\sum_{i=1}^{n} \alpha_i(u, q_i)(R_\lambda q_i, q_j)$$

or

$$\sum_{i=1}^{n} \{\delta_{ij} + \alpha_i(R_\lambda q_i, q_j)\}(u, q_i) = 0.$$

Not all $(u, q_i) = 0$, since this would imply, contrary to our assumption, that λ is an eigenvalue of A. Therefore, by the same analysis as given in Section 4.3, λ is a nonpersistent eigenvalue of (4) if and only if

$$V_{0n}(\lambda) = \det\{\delta_{ij} + \alpha_i(R_\lambda q_i, q_j)\} = 0 \qquad (i, j = 1, 2, \ldots, n). \tag{6}$$

Moreover, the multiplicity of λ as an eigenvalue of (4) is equal to the nullity of

$$\{\delta_{ij} + \alpha_i(R_\lambda q_i, q_j)\} \qquad (i, j = 1, 2, \ldots, n). \tag{7}$$

This determinant (6) is called the *modified Weinstein determinant* or the *Weinstein–Aronszajn determinant*, or, for short, the *WA determinant*; see (4.3.4) for comparison. The matrix (7) appears to be nonsymmetric but it can be replaced by a symmetric matrix in several ways. The new determinant may differ from (7) by a constant factor. One way to obtain a symmetric matrix is to introduce new vectors $v_i = \alpha_i^{1/2} q_i (i = 1, 2, \ldots, n)$, in which case the matrix in (7) becomes

$$\det\{\delta_{ij} + (R_\lambda v_i, v_j)\}.$$

In fact, multiplying the ith row by $\alpha_i^{1/2}$ $(i = 1, 2, \ldots, n)$ and the jth column by $\overline{\alpha_j^{-1/2}}$ $(j = 1, 2, \ldots, n)$, we have

$$\begin{aligned}\det\{\delta_{ij} + (R_\lambda v_i, v_j)\} &= \pm\det\{\delta_{ij} + \alpha_i^{1/2}[\overline{\alpha_j^{-1/2}}](R_\lambda v_i, v_j)\} \\ &= \det\{\delta_{ij} + \alpha_i^{1/2}[\overline{\alpha_j^{-1/2}}](R_\lambda \alpha_i^{1/2} q_i, \alpha_j^{1/2} q_j) \\ &= \det\{\delta_{ij} + \alpha_i (R_\lambda q_i, q_j)\}.\end{aligned}$$

If all α_i are positive, the vectors v_i are mutually orthogonal vectors having magnitude $\alpha_i^{1/2}$. Another procedure will be discussed in Section 9.2.

For completeness, let us add that if the q_i were arbitrary vectors [see (1)], we would obtain instead of (6) the determinant

$$\det\left\{\delta_{ik} + \sum_{j=1}^{n} \alpha_{ij}(R_\lambda q_j, q_k)\right\} \qquad (i, k = 1, 2, \ldots, n),$$

which differs from (6) by a constant factor (compare Section 9.2).

3. A Solution of the Second Type

This section parallels Section 4.8. The results have never been used for numerical computations in the second type but have great theoretical interest. We begin with the problem

$$Au - \lambda u = -\alpha(u, q)q, \tag{1}$$

in which we have put $\alpha_1 = \alpha$ and $q = q_1$ for simplicity. Suppose that λ is an eigenvalue of A, say λ_k. Again we let m and M be the smallest and largest indices, respectively, such that $\lambda_m = \lambda_k = \lambda_M$. If we consider the function $V_{01}(\lambda) = 1 + \alpha(R_\lambda q, q)$, then there are three possibilities.

Case A. If $V_{01}(\lambda_k) = \infty$, the vector q is not orthogonal to the eigenspace of λ_k. In this case, the solutions of (1) are given by

$$u = \sum_{j=m}^{M} \beta_j u_j,$$

where the β_j are chosen so that $(u, q) = 0$. In this way, the multiplicity of λ_k is diminished by one. As usual, multiplicity zero means that λ_k is not an eigenvalue of (1).

CASE B. If $0 < |V_{01}(\lambda_k)| < \infty$, then the vector q must be orthogonal to each u_j $(j = m, m+1, \ldots, k, \ldots, M)$. The most general eigenfunction has the form

$$u = \beta_0 R'_{\lambda_k} q + \sum_{j=m}^{M} \beta_j u_j, \tag{2}$$

where $\beta_m, \beta_{m+1}, \ldots, \beta_M$ are arbitrary scalars, but β_0 is a linear functional of u and cannot be considered as arbitrary. Putting this vector (2) into (1), we see that β_0 must satisfy

$$\beta_0 q = -\alpha\beta_0(R'_{\lambda_k} q, q)q$$

or

$$\beta_0[1 + \alpha(R'_{\lambda_k} q, q)]q = 0,$$

which, in view of the fact that $V_{01}(\lambda_k) \neq 0$, is possible only if $\beta_0 = 0$. Therefore, the only independent solutions are $u_m, u_{m+1}, \ldots, u_M$, and the multiplicity is preserved.

CASE C. If $V_{01}(\lambda_k) = 0$, then again the vector q must be orthogonal to each u_j $(j = m, m+1, \ldots, k, \ldots, M)$. Let $u = \alpha R'_{\lambda_k} q$. Since $V_{01}(\lambda_k) = 1 + (u, q) = 0$, we have $(u, q) = -1$, so that u is a nontrivial solution of (1). Let us note that if we were to write (1) as

$$Au - \lambda u = \beta_0 q$$

and treat β_0 as an "arbitrary scalar," then we would obtain a solution $u = \beta_0 R'_{\lambda_k} q$, but we would not know that the β_0 is not identically zero for the actual solution. (For nonpersistent eigenvalues, it is obvious that $\beta_0 \neq 0$, a fact emphasized by Weinstein [W9].) Besides $\alpha R'_{\lambda_k} q$, each u_j $(j = m, m+1, \ldots, k, \ldots, M)$ is a solution of (1) and since $R'_{\lambda_k} q$ is orthogonal to the eigenspace of λ_k, the vectors $R'_{\lambda_k} q, u_m, u_{m+1}, \ldots, u_M$, are linearly independent solutions of (1). Therefore, the multiplicity of λ_k is increased by one. All the above cases for persistent as well as nonpersistent eigenvalues may be condensed into the following rule.

Theorem 1. (*Aronszajn's rule*). *Let $\mu_0(\lambda)$ and $\mu_1(\lambda)$ denote the multiplicities of λ as an eigenvalue of A and A_1, respectively. Let $\omega_{01}(\lambda)$ be the order of the function $V_{01}(\xi) = 1 + \alpha(R_\xi q, q)$ at the point $\xi = \lambda$. Then $\mu_1(\lambda) = \mu_0(\lambda) + \omega_{01}(\lambda)$ for all $\lambda < \lambda_\infty$.*

The eigenvalue equation for an operator perturbed by a bounded symmetric operator of rank r can be written as

$$A_r u = Au + \sum_{i=1}^{r} \alpha_i(u, q_i)q_i = \lambda u, \tag{3}$$

where $\alpha_i \neq 0$ $(i = 1, 2, \ldots, r)$ and $(q_i, q_j) = \delta_{ij}$ $(i, j = 1, 2, \ldots, r)$.

In order to solve eigenvalue problem (3), we solve a sequence of problems in which each operator differs from the preceeding one by an operator of rank one. By adding successive perturbations of rank one, we obtain, for each of the problems

$$A_k u - \lambda u = -\alpha_{k+1}(u, q_{k+1})q_{k+1} \qquad (k = 0, 1, \ldots, r-1), \tag{4}$$

corresponding functions (where $R_\lambda{}^k = [A_k - \lambda I]^{-1}$)

$$V_{k,k+1}(\lambda) = 1 + \alpha_{k+1}(R_\lambda{}^k q_{k+1}, q_{k+1}) \qquad (k = 0, 1, \ldots, r-1), \tag{5}$$

which have the property that the difference between the multiplicity of λ as an eigenvalue of A_k and A_{k+1} is given by the order of $V_{k,k+1}$ at λ (see Theorem 1).

However, we wish to solve problem (3) solely in terms of the resolvent of A and not the resolvents of $A_1, A_2, \ldots, A_{r-1}$. In order to accomplish this, we use the following decomposition.

Theorem 2.

$$V_{01}(\lambda) = \det\{\delta_{ik} + \alpha_i(R_\lambda q_i, q_k)\} = \prod_{j=0}^{r-1} V_{j,j+1}(\lambda) \qquad (i, k = 1, 2, \ldots, r). \tag{6}$$

PROOF. Let j be a given index $(2 \leq j \leq r)$. We shall show that $V_{0j}(\lambda) = V_{0,j-1}(\lambda)V_{j-1,j}(\lambda)$, from which formula (6) will follow immediately by induction. Let v be a solution of the problem

$$Av + \sum_{i=1}^{j-1} \alpha_i(v, q_i)q_i - \lambda v = \alpha_j q_j. \tag{7}$$

If we assume for the moment that λ is not an eigenvalue of either A or A_{j-1}, we have $v = \alpha_j R_\lambda^{j-1} q_j$. Problem (7) could also be written as

$$\begin{aligned} Av - \lambda v &= -\sum_{i=1}^{j-1} \alpha_i(v, q_i)q_i + \alpha_j q_j \\ &= -\sum_{i=1}^{j-1} \alpha_j \alpha_i(R_\lambda^{j-1} q_j, q_i)q_i + \alpha_j q_j, \end{aligned}$$

in which form we see that

$$\alpha_j R_\lambda^{j-1} q_j = v = -\sum_{i=1}^{j-1} \alpha_j \alpha_i(R_\lambda^{j-1} q_j, q_i)R_\lambda q_i + \alpha_j R_\lambda q_j.$$

Therefore, we have (since $\alpha_j \neq 0$)

$$\sum_{i=1}^{j-1} \alpha_i (R_\lambda^{j-1} q_j, q_i) R_\lambda q_i + R_\lambda^{j-1} q_j = R_\lambda q_j. \tag{8}$$

Taking the inner product of both sides of (8) with $q_1, q_2, \ldots, q_j$, we obtain the set of equations

$$\sum_{i=1}^{j-1} \alpha_i (R_\lambda^{j-1} q_j, q_i)(R_\lambda q_i, q_k) + (R_\lambda^{j-1} q_j, q_k)$$
$$= (R_\lambda q_j, q_k) \qquad (k = 1, 2, \ldots, j)$$

or equivalently

$$\sum_{i=1}^{j-1} \{\delta_{ik} + \alpha_i (R_\lambda q_i, q_k)\}(R_\lambda^{j-1} q_j, q_i)$$
$$= (R_\lambda q_j, q_k) \qquad (k = 1, 2, \ldots, j-1) \tag{9}$$

and

$$\sum_{i=1}^{j-1} \alpha_i (R_\lambda q_i, q_j)(R_\lambda^{j-1} q_j, q_i) = (R_\lambda q_j, q_j) - (R_\lambda^{j-1} q_j, q_j). \tag{10}$$

We can now write $(i, k = 1, 2, \ldots, j-1)$

$$V_{0j}(\lambda) = \begin{vmatrix} \delta_{ik} + \alpha_i (R_\lambda q_i, q_k) & \alpha_i (R_\lambda q_i, q_j) \\ \alpha_j (R_\lambda q_j, q_k) & 1 + \alpha_j (R_\lambda q_j, q_j) \end{vmatrix}.$$

Subtracting $\alpha_j (R_\lambda^{j-1} q_j, q_i)$ times the ith row from the jth row $(i = 1, 2, \ldots, j-1)$ and using the system of equations (9) and (10), we see that

$$V_{0j}(\lambda) = \begin{vmatrix} \delta_{ik} + \alpha_i (R_\lambda q_i, q_k) & \alpha_i (R_\lambda q_i, q_j) \\ 0 & 1 + \alpha_j (R_\lambda^{j-1} q_j, q_j) \end{vmatrix}$$
$$= V_{0, j-1}(\lambda) V_{j-1, j}(\lambda),$$

which yields (6) for λ not on the spectrum of $A, A_1, \ldots,$ or A_{j-1}. Since each of the functions $V_{01}(\lambda), V_{12}(\lambda), \ldots, V_{j-1, j}(\lambda)$ is meromorphic on the lower part of the spectrum of $A, A_1, \ldots, A_{j-1}$, respectively (see Section 4.8), decomposition (6) remains valid even if λ is an isolated eigenvalue of finite multiplicity of $A, A_1, \ldots,$ or A_{j-1}. The notation $V_{0r}(\lambda)$ is used to signify that this determinant links the spectrum of the operator A_0 with that of A_r.

Decomposition (6) is useful in several ways. One way is to combine it with Theorem 1 to yield the following formulation of *Aronszajn's rule.*

Theorem 3. *Let $\mu_r(\lambda)$ denote the multiplicity of λ as an eigenvalue of A_r and let $\omega_{0r}(\lambda)$ be the order of $V_{0r}(\xi)$ at $\xi = \lambda$. Then,*

$$\mu_r(\lambda) = \mu_0(\lambda) + \omega_{0r}(\lambda). \tag{11}$$

We see that this rule concerning the second type (3) coincides with rule (4.8.1) for the first type given by Aronszajn in 1948 [A1].

As we have seen in Section 4.9 there are difficulties in applying Aronszajn's rule to numerical computations in intermediate problems of higher order. Instead Bazley, Fox, and others used other procedures, some of which have their roots in Weinstein's earliest work (see Section 4.4), and will be discussed in the following sections.

Another application of decomposition (6) will be given in Section 9.2.

4. Bazley's Special Choice

In the second type of intermediate problems, especially in the case of Schrödinger's equation, the computational difficulties when the vectors p_i are left general were such that for about ten years no numerical applications of the second type of problems were given and problems of the second type seemed to be confined to their own theoretical content. A major breakthrough was achieved by Bazley [Bl–3], who introduced a special choice of the vectors p_i, thereby solving numerically some important problems of the second type. We shall see that Bazley's special choice in some sense corresponds to the distinguished choice of Weinstein's (Section 4.4), so that the chronological development here was the reverse of that in the first type of problems.

Bazley assumes that A has at least n eigenvalues whose corresponding eigenvectors belong to the range of B. This assumption is without loss of generality since we can reformulate the given problem as

$$(\tilde{A} + \tilde{B})u = (A - \gamma I + B + \gamma I)u = \lambda u,$$

where $\gamma > 0$. Since $\tilde{B} = B + \gamma I$ is bounded away from zero, every eigenvector of $\tilde{A} = A - \gamma I$ is in the range of $\tilde{B}$.

For simplicity, we shall consider $\lambda_1^{(0)}, \lambda_2^{(0)}, \ldots, \lambda_n^{(0)}$, but actually we could have considered, say, $\lambda_5^{(0)}, \lambda_7^{(0)}, \lambda_{10}^{(0)}$, etc. We choose vectors p_i such that

$$Bp_i = u_i^{(0)} \qquad (i = 1, 2, \ldots, n). \tag{1}$$

In this way, Eq. (1.4) becomes

$$A_n u = Au + \sum_{i=1}^{n} \sum_{j=1}^{n} \beta_{ij}(u, u_i^{(0)})u_j^{(0)} = \lambda u. \tag{2}$$

We can obtain all eigenvalues and eigenvectors of (2) in an easy way. Our main tool will be the decomposition of A and A_n into orthogonal parts and an application of Theorem A.4. Let us denote by U_n the orthogonal projection operator onto $\mathfrak{U}_n = sp\{u_1^{(0)}, u_2^{(0)}, \ldots, u_n^{(0)}\}$ and let V_n be the orthogonal projection operator onto $\mathfrak{V}_n = \mathfrak{U}_n^{\perp}$, so that $U_n + V_n = I$. Then, we can write

$$A = U_n A + V_n A$$

and

$$A_n = U_n A_n + V_n A_n.$$

Moreover, since the $u_i^{(0)}$ $(i = 1, 2, \ldots, n)$ are eigenvectors of A, the projectors U_n and V_n commute with A. Also, in view of (2), we see that U_n and V_n commute with A_n. Therefore, we have

$$\begin{aligned} A &= U_n A + V_n A \\ &= U_n U_n A + V_n V_n A \\ &= U_n A U_n + V_n A V_n \end{aligned} \tag{3}$$

and similarly

$$A_n = U_n A_n U_n + V_n A_n V_n. \tag{4}$$

In order to apply Theorem A.4, we consider the restrictions (compare Section 3.2) of A to the *invariant* subspaces $\mathfrak{U}_n$ and $\mathfrak{V}_n$. We write these restrictions as $A|_{\mathfrak{U}_n}$ and $A|_{\mathfrak{V}_n}$. Then, the spectrum of A separates as

$$\sigma(A) = \sigma(A|_{\mathfrak{U}_n}) \cup \sigma(A|_{\mathfrak{V}_n}).$$

Similarly, the spectrum of A_n separates as

$$\sigma(A_n) = \sigma(A_n|_{\mathfrak{U}_n}) \cup \sigma(A_n|_{\mathfrak{V}_n}).$$

Before proceeding further, let us note that in Section 3.2 we considered the *part of A in* $\mathfrak{Q}$ where $\mathfrak{Q}$ was not necessarily an invariant subspace for A, in which case Theorem A.4 on the separation of the spectrum *does not apply*.

We now note that $V_n u_j^{(0)} = 0$, $(j = 1, 2, \ldots, n)$, so that

$$A|_{\mathfrak{V}_n} = A_n|_{\mathfrak{V}_n}.$$

Therefore, the only difference between the spectra of A and A_n is the difference between $\sigma(A|_{\mathfrak{U}_n})$ and $\sigma(A_n|_{\mathfrak{U}_n})$. All other eigenvalues are persistent. By inspection, we see that

$$\sigma(A|_{\mathfrak{U}_n}) = \{\lambda_1^{(0)}, \lambda_2^{(0)}, \ldots, \lambda_n^{(0)}\}.$$

Since $A_n|_{\mathfrak{U}_n}$ is an operator from the n-dimensional space $\mathfrak{U}_n$ to itself, the spectrum consists of n (not necessarily distinct) eigenvalues which are given by the characteristic roots of a matrix representing the operator $A_n|_{\mathfrak{U}_n}$.

We now derive such a matrix. Since $\mathfrak{U}_n = \text{sp}\{u_1^{(0)}, u_2^{(0)}, \ldots, u_n^{(0)}\}$, a typical vector $v \in \mathfrak{U}_n$ may be written as

$$v = \sum_{h=1}^{n} \alpha_h u_h^{(0)}. \tag{5}$$

If v is an eigenvector of $A_n|_{\mathfrak{U}_n}$ corresponding to λ, we must have

$$Av + \sum_{i,j=1}^{n} \beta_{ij}(v, u_i^{(0)})u_j^{(0)} = \lambda v,$$

which means, in view of (5), that

$$\sum_{h=1}^{n} \alpha_h \lambda_h^{(0)} u_h^{(0)} + \sum_{h,j=1}^{n} \beta_{hj} \alpha_h u_j^{(0)} = \lambda \sum_{h=1}^{n} \alpha_h u_h^{(0)}. \tag{6}$$

Since the $u_i^{(0)}$ are independent, a nontrivial solution of (6) exists if and only if

$$\det\{(\lambda_h^{(0)} - \lambda)\,\delta_{hj} + \beta_{hj}\} = 0 \qquad (h, j = 1, 2, \ldots, n). \tag{7}$$

The n roots $\mu_1 \leq \mu_2 \leq \cdots \leq \mu_n$ of this determinant give n eigenvalues of A_n some or all of which *might* be nonpersistent. By ordering the eigenvalues μ_i and the additional persistent eigenvalues $\lambda_{n+1}^{(0)} \leq \lambda_{n+2}^{(0)} \leq \cdots$ in a single nondecreasing sequence

$$\lambda_1^{(n)} \leq \lambda_2^{(n)} \leq \cdots \tag{8}$$

we obtain all eigenvalues in the lower part of the spectrum of A_n. This part of the spectrum yields lower bounds satisfying the inequalities

$$\lambda_i^{(0)} \leq \lambda_i^{(n)} \leq \lambda_i^{(\infty)} \qquad (i = 1, 2, \ldots).$$

The derivation given above avoids the discussion of the continuous spectrum investigated in [B3, p. 297].

The special choice considered here corresponds to Weinstein's distinguished choice in the first type of intermediate problems (Section 4.4). In fact, in the special choice we have

$$(Bp_i, u_j^{(0)}) = (u_i^{(0)}, u_j^{(0)}) = \delta_{ij} \qquad (i, j = 1, 2, \ldots, n),$$

while in the distinguished choice we have

$$(p^i, u_j) = \delta_{ij} \qquad (i, j = 1, 2, \ldots, n).$$

There would be a complete parallel if it were possible in a given problem of the first type to take some $p^i = u_i^{(0)}$.

As in the case of the distinguished choice, we have the following theorem.

Theorem 1. *Let n be an index such that $\lambda_n^{(0)} < \lambda_{n+1}^{(0)}$. Let $p_1, p_2, \ldots, p_n$ be the special choice* (1). *Then, the first n eigenvalues of the nth intermediate problem* (2) *satisfy the strict inequalities*

$$\lambda_i^{(0)} < \lambda_i^{(n)} \qquad (i = 1, 2, \ldots, n) \tag{9}$$

and, therefore, have been improved.

Proof. The roots

$$\mu_1 \leq \mu_2 \cdots \leq \mu_n$$

of determinant (7) are the eigenvalues of the matrix [see (2.3.10)]

$$M = \{\lambda_h^{(0)} \delta_{hj} + \beta_{hj}\} \qquad (h, j = 1, 2, \ldots, n),$$

while, of course, $\lambda_1^{(0)}, \lambda_2^{(0)}, \ldots, \lambda_n^{(0)}$ are the eigenvalues of the matrix

$$L = \{\lambda_h^{(0)} \delta_{hj}\} \qquad (h, j = 1, 2, \ldots, n).$$

Since $\{\beta_{hj}\}$ is the inverse of the Gram matrix [see (1.4)] and the Gram matrix is positive-definite for independent vectors [14, p. 16], it follows that $\{\beta_{hj}\}$ is itself positive-definite. Therefore, M strictly dominates L, $L < M$, and by the principle of monotonicity (2.5.3) we have

$$\lambda_i^{(0)} < \mu_i \qquad (i = 1, 2, \ldots, n). \tag{10}$$

Forming the single sequence (8) of all eigenvalues, we see that, if the index $i \leq n$ is such that $\mu_i \leq \lambda_{n+1}^{(0)}$, then

$$\lambda_i^{(n)} = \mu_i,$$

so that by (10) we have

$$\lambda_i^{(0)} < \lambda_i^{(n)}.$$

On the other hand, if the index $i \leq n$ is such that $\mu_i > \lambda_{n+1}^{(0)}$, then either $\lambda_i^{(n)}$ will be equal to some $\mu_j > \lambda_{n+1}^{(0)}$ or else it will be equal to some $\lambda_{n+h}^{(0)}$, $h \geq 1$. In these cases, we have

$$\lambda_i^{(0)} < \lambda_{n+1}^{(0)} \leq \lambda_i^{(n)},$$

and therefore (9) holds.

In some numerical applications, the inversion $Bp_i = u_i^{(0)}$ may not be easy, but an appropriate choice of p_i yields equations of the type

$$Bp_i = \sum_{j=1}^{N(k)} \eta_{ij} u_j^{(0)} \qquad (i = 1, 2, \ldots, k) \tag{11}$$

(see Bazley [B3] and Bazley and Fox [BF2]). The choice (11) is sometimes called the *generalized special choice*. Obviously, intermediate problems here have essentially the same structure as in the special choice (1).

5. The General Choice and Truncation

The method of special choice is not always numerically feasible. Therefore, it is of importance to consider a rule for a general choice of the p_i.

The nonpersistent eigenvalues have already been discussed in Section 2. For persistent eigenvalues, we have the following rule, which parallels the results of Section 4.7.

Theorem 1. *The multiplicity of a persistent eigenvalue* λ_* *is given by the nullity of the matrix*

$$\begin{array}{cc} n & \mu \end{array}$$
$$\begin{bmatrix} (p_i + R'_{\lambda_*} Bp_i, Bp_k) & (u^h, Bp_k) \\ (Bp_i, u^h) & 0 \end{bmatrix} \qquad \begin{pmatrix} i, k = 1, 2, \ldots, n \\ h = 1, 2, \ldots, \mu \end{pmatrix},$$

where again $u^1, u^2, \ldots, u^\mu$ *are the eigenvectors corresponding to* λ_* *in the base problem.*

The advantage of this method is that every element is finite. As mentioned in Section 4.10, we may still have difficulties here in the numerical computations, which can be avoided by the procedure of truncation. The truncation base operator here has the same form as in the first type, namely

$$\begin{aligned} T_N u &= \sum_{i=1}^{N} \lambda_i (u, u_i^{(0)}) u_i^{(0)} + \lambda_{N+1} \int_{\lambda_{N+1}}^{\infty} dE_\lambda u \\ &= \sum_{i=1}^{N} \lambda_i (u, u_i^{(0)}) u_i^{(0)} + \lambda_{N+1} \left[u - \sum_{i=1}^{N} (u, u_i^{(0)}) u_i^{(0)} \right]. \end{aligned} \tag{1}$$

In the same way as in the first type, intermediate problems of the second type can be formed relative to T_N for *any* choice of the p_i.

The special choice and later the generalized special choice and truncation were applied to the important problem of lower bounds for the first 25 eigenvalues of the Mathieu equation

$$\begin{gathered} -(d^2u/dx^2) + (s\cos^2 x)u = \lambda u, \qquad 0 \le x \le \pi, \\ u'(0) = u'(\pi) = 0, \\ u(\tfrac{1}{2}\pi - x) = u(\tfrac{1}{2}\pi + x), \end{gathered} \tag{2}$$

for $s = 1, 2, 4, 8, 16, 32$, in all 150 values. These results, compared with the Rayleigh–Ritz values, are remarkably accurate (see Tables B.21–27). In (2) the base operator is given by $Au = -u''$.

6. The Existence of a Base Problem for a Compact Operator

In the first type of intermediate problems, the base problem is often obtained by the removal of constraints. The existence of a base problem in the second type of problems is by no means evident. The question of the existence of a base problem for a given problem is a fundamental point of the theory and determines the form of intermediate problems.

Let us now formulate Weinstein's result [W25].

Theorem 1. *Given a compact operator K, there exists a family of explicity solvable, noncompact operators each of which provides lower (upper) bounds for a finite but arbitrary number of the negative (positive) eigenvalues of K and each of which can be used as a base problem.*

PROOF. It will suffice to prove this assertion for the negative eigenvalues. We consider any given complete orthonormal sequence $v_1, v_2, \ldots$ in $\mathfrak{H}$, we put $\alpha_{ij} = (Kv_i, v_j)$ $(i, j = 1, 2, \ldots)$, and define degenerate operators K_n by

$$K_n u = \sum_{i,j=1}^{n} (u, v_i)\alpha_{ij} v_j \qquad (n = 1, 2, \ldots). \tag{1}$$

Since the K_n are degenerate, as in Section 2.3, the (nontrivial) eigenvalues $\Lambda_1 \le \Lambda_2 \le \cdots \le \Lambda_n$ are given by the roots of the equation

$$\det\{\alpha_{ij} - \Lambda\, \delta_{ij}\} = 0 \qquad (i, j = 1, 2, \ldots, n). \tag{2}$$

We denote by $u_1^{(0)}, u_2^{(0)}, \ldots, u_n^{(0)}$ the eigenvectors corresponding to $\Lambda_1, \Lambda_2, \ldots, \Lambda_n$. We assume without loss of generality that $\Lambda_j < 0$ $(j = 1, 2, \ldots, n)$, since we can eliminate from the degenerate operator (1) any eigenvector corresponding to a positive or zero root of (2). Clearly, $u_{n+1}^{(0)} = v_{n+1}$, $u_{n+2}^{(0)} = v_{n+2}, \ldots$ are eigenvectors corresponding to the (trivial) eigenvalue $\Lambda = 0$ of K_n, which is generally of infinite multiplicity.

We recall the well-known result that $\|Ku - K_n u\| \to 0$ uniformly in u as $n \to \infty$ (see [26, p. 204]). Therefore, for all sufficiently large n, we have $\|K - K_n\| < \|K\|$, where $\|\cdot\|$ denotes the operator norm. Let γ_n be any (positive) number such that

$$\|K - K_n\| < \gamma_n < \|K\|.$$

We now write

$$K = K_n - \gamma_n I + [K - K_n] + \gamma_n I.$$

Since $B_n = K - K_n + \gamma_n I$ is strictly positive, the operator $A_n = K_n - \gamma_n I$ is dominated by K. Moreover, the *noncompact* operator A_n has known eigenvalues

$$\lambda_1^{(0)} = \Lambda_1 - \gamma_n \le \lambda_2^{(0)} = \Lambda_2 - \gamma_n, \ldots, \quad \lambda_n^{(0)} = \Lambda_n - \gamma_n,$$
$$\lambda_{n+1}^{(0)} = -\gamma_n, \quad \lambda_{n+2}^{(0)} = -\gamma_n, \ldots.$$

The spectral formula for A_n is

$$A_n u = \sum_{i=1}^{n} (\Lambda_i - \gamma_n)(u, u_i^{(0)})u_i^{(0)} - \gamma_n\left(u - \sum_{i=1}^{n} (u, u_i^{(0)})u_i^{(0)}\right).$$

From this representation, it is clear that A_n is of the same type as a truncated operator in that it has a finite number of eigenvalues of finite multiplicity followed by an eigenvalue of infinite multiplicity. It must be mentioned, however, that, unlike in Sections 4.10 and 5, there is no "original" base operator, the construction of which is our problem here. Indeed, the known operator A_n can be used as a base problem for our operator K, thus proving Theorem 1. It is interesting to note that the computational idea of truncation is used here for an existence theorem.

In applying the above construction, the following choices are available.

Case A. One can fix an index n and retain A_n as a base operator, then increase the order of the intermediate problem in order to obtain improved bounds for n eigenvalues of K.

Case B. One can increase n, thereby changing the base problem, then apply an intermediate problem of a given order to obtain bounds for more eigenvalues of K.

It may seem unusual that our result yields a *noncompact* base operator for any compact operator. This is especially remarkable in the case of an integral operator where one would expect the base operator to be another integral operator. Of course, this does not preclude the fact that some compact operators have compact base operators, for instance, if the given operator admits the obvious decomposition $A + B$ as in the previous sections. Also, in a recent paper, Stakgold [S4] showed that it is always possible to construct a compact base operator and construct intermediate problems of the *first type*, provided the given operator is a compact, selfadjoint, integral operator generated by a difference kernel.

Let us note that the proof of convergence in the present case is very simple due to the fact that the base problem is constructed directly from the given problem. In fact, we have

$$\|K - A_n\| = \|K - K_n + \gamma_n I\| \leq \|K - K_n\| + \gamma_n .$$

For sufficiently large n, we can take

$$\gamma_n = [1 + (1/n)]\|K - K_n\|.$$

Therefore, we have

$$\|K - A_n\| \leq [2 + (1/n)]\|K - K_n\|.$$

Since $\|K - K_n\| \to 0$ as $n \to \infty$, we have $\|K - A_n\| \to 0$ as $n \to \infty$, so that the eigenvalues of the base problems A_n converge to the unknown eigenvalues. Since the eigenvalues of the intermediate problems lie between those of the base problem and those of the given problem, they must also converge to the unknown eigenvalues.

Let us note in passing that, after the degenerate operator K_n has been obtained, the procedure of intermediate problems, even in the case of integral equations, is completely different from the classical method of E. Schmidt.

The existence of a base problem for all operators of class $\mathscr{S}$ has not yet been established. Fortunately, base operators have been found explicitly for many problems in quantum mechanics as well as classical differential equations (see Appendix B).

7. The Spectrum of the Helium Atom

One of the most outstanding numerical applications of the second type of intermediate problem is Bazley's determination of lower bounds for the energy levels (eigenvalues) of the helium atom. We first give some theoretical results due to Kato [K2–3] and Zislin and Sigalov [Z4, ZSI–2] for the Schrödinger equation of the helium atom.

The helium atom consists of a nucleus (of infinite mass) and two electrons. Letting the nucleus be the origin, we denote by (x_1, y_1, z_1) and (x_2, y_2, z_2) the coordinates of the two electrons. The Schrödinger operator for helium which we consider here is

$$Hu = -\tfrac{1}{2}\Delta_1 u - \tfrac{1}{2}\Delta_2 u - (2/r_1)u - (2/r_2)u + (1/r_{12})u, \tag{1}$$

where Δ_i is the Laplacian in the coordinates r_i,

$$r_i = (x_i^2 + y_i^2 + z_i^2)^{1/2} \qquad (i = 1, 2),$$
$$r_{12} = [(x_2 - x_1)^2 + (y_2 - y_1)^2 + (z_2 - z_1)^2]^{1/2}.$$

While the domain of definition of H from the point of view of the physicist was originally only vaguely defined, Kato [*K*2] considered the Hilbert space of square-integrable functions over six-dimensional coordinate space and proved that H admits there a unique selfadjoint extension. This is exactly what is meant by saying that H is essentially selfadjoint. In other words, the closure of H, which we denote by $\tilde{H}$, is selfadjoint [19, p. 269]. Of course, $\tilde{H}$ is no longer a differential operator in the usual sense, but it reduces to the differential operator for sufficiently regular functions.

While, by the general spectral theorem, $\tilde{H}$ admits the representation

$$\tilde{H} = \lambda \int_{-\infty}^{\infty} dE_\lambda, \tag{2}$$

it is by no means a foregone conclusion that $\tilde{H}$ is of class $\mathscr{S}$. It is therefore significant that Kato [K2] showed that the spectrum of $\tilde{H}$ begins with isolated eigenvalues, each of finite multiplicity. In order to present Kato's theory, we have to discuss in some detail the fundamental properties of the spectral resolution (2).

Let $\mathfrak{M}_\lambda$ denote the range of the projector E_λ, so that $E_\lambda \tilde{H} = \mathfrak{M}_\lambda$. For any $f \in \mathfrak{H}$, $(E_\lambda f, f)$ is a monotonically increasing function of λ. We say for short that E_λ is monotonically increasing. We recall that E_λ is constant on the resolvent set.

Suppose that $\alpha < \beta$. Let $P_{\alpha,\beta} = E_\beta - E_\alpha$ and $\mathfrak{M}_{\alpha,\beta} = P_{\alpha,\beta}\mathfrak{H}$. Then, $\mathfrak{M}_{\alpha,\beta} = \mathfrak{M}_\beta \ominus \mathfrak{M}_\alpha$, that is, $\mathfrak{M}_{\alpha,\beta}$ is the orthogonal complement of $\mathfrak{M}_\alpha$ in $\mathfrak{M}_\beta$.

If E_λ is constant on the interval $-\infty < \lambda \leq \mu$, then $\mathfrak{M}_\mu$ is obviously the zero space, $\mathfrak{M}_\mu = \{0\}$. If E_λ changes continuously on any subinterval of $(-\infty, \mu]$, say on (α_0, β_0), then $\dim \mathfrak{M}_\mu = \infty$. In fact, by choosing an infinite sequence of points $\alpha_0 < \alpha_1 < \alpha_2 < \cdots < \beta_0$, we have, for every integer $n = 1, 2, \ldots$,

$$\mathfrak{M}_{\alpha_n} = \mathfrak{M}_{\alpha_0} \oplus \mathfrak{M}_{\alpha_0, \alpha_1} \oplus \cdots \oplus \mathfrak{M}_{\alpha_{n-1}, \alpha_n} \tag{3}$$

and

$$\mathfrak{M}_\mu = \mathfrak{M}_{\alpha_n} \oplus \mathfrak{M}_{\alpha_n, \mu}. \tag{4}$$

Since each of the subspaces $\mathfrak{M}_{\alpha_i, \alpha_{i+1}}$ $(i = 0, 1, \ldots, n-1)$ is nonzero, we have

$$1 \leq \dim \mathfrak{M}_{\alpha_i, \alpha_{i+1}} \qquad (i = 0, 1, \ldots, n-1). \tag{5}$$

Therefore, from (3)–(5), we have

$$n \leq \dim \mathfrak{M}_{\alpha_n} \leq \dim \mathfrak{M}_\mu. \tag{6}$$

As (6) is true for any $n = 1, 2, \ldots$, we must have $\dim \mathfrak{M}_\mu = \infty$. Similarly, if E_λ has an infinite number of discontinuities on $-\infty < \lambda \leq \mu$ or if E_λ has a discontinuity at a point α, $-\infty < \alpha \leq \mu$, such that $\dim \mathfrak{M}_{\alpha-0} = \infty$, then of course $\dim \mathfrak{M}_\mu = \infty$.

The above analysis can be summarized in the following two statements.

a. If there are only resolvent points in $(-\infty, \mu]$, then $\dim \mathfrak{M}_\mu = 0$.

b. If there exists a point of the essential spectrum in $(-\infty, \mu)$, then $\dim \mathfrak{M}_\mu = \infty$.

Using a and b, we can give the following criterion for operators of class $\mathscr{S}$.

Theorem 1. *A selfadjoint operator is in class $\mathscr{S}$ if and only if there exist points μ_1 and μ_2 $(\mu_1 < \mu_2)$ such that*

$$0 < \dim \mathfrak{M}_{\mu_1} < \infty \tag{7}$$

and

$$0 < \dim \mathfrak{M}_{\mu_2} < \infty. \tag{8}$$

PROOF. If $\tilde{H} \in \mathscr{S}$, we can put $\mu_1 = \lambda_1$, the first eigenvalue, and μ_2 any point in the interval $(\lambda_1, \lambda_\infty)$, from which (7) and (8) follow immediately. On the other hand, suppose (7) and (8) hold. Applying b at the point $\mu = \mu_2$, we see from (8) that μ_1 cannot be in the essential spectrum. Applying a at the point $\mu = \mu_1$, we see from (7) that the interval $(-\infty, \mu_1]$ contains a finite, nonzero number of isolated eigenvalues of finite multiplicity and no other spectral points, which concludes the proof of the theorem.

REMARK. If we were to postulate only the existence of one point, say μ_2, such that (8) holds, then μ_2 could be the lowest point in the spectrum while at the same time μ_2 is an eigenvalue of finite multiplicity and *also* in the essential spectrum. In this case, there would be no isolated point eigenvalues at the beginning of the spectrum. For instance, this could occur for a positive compact operator at the point $\mu_2 = 0$.

In order to apply this criterion to $\tilde{H}$, we make use of the following two general lemmas.

Lemma 1. *Let H be any selfadjoint operator with domain $\mathfrak{D}$, let E_λ be the corresponding resolution of the identity, and let $\mathfrak{M}_\lambda = E_\lambda \mathfrak{H}$. If there is an N-dimensional subspace $\mathfrak{B} \subset \mathfrak{D}$ such that*

$$(Hf, f) \leq \mu(f, f) \tag{9}$$

for every $f \in \mathfrak{B}$, then $\dim \mathfrak{M}_\mu \geq N$.

PROOF. Suppose, contrary to the assertion that $\dim \mathfrak{M}_\mu < N$. Then there exists (compare with Section 3.3) $f \in \mathfrak{B}$, $(f, f) = 1$, such that f is orthogonal to all vectors in $\mathfrak{M}_\mu$, that is, $E_\mu f = 0$ (compare with Section 3.1). From the spectral formula, we have

$$(Hf, f) = \int_{-\infty}^{\infty} \lambda \, d(E_\lambda f, f) = \int_{\mu+0}^{\infty} \lambda \, d(E_\lambda f, f) \geq \mu(f, f). \tag{10}$$

We suppose that $(Hf, f)/(f, f) = \mu$. Then, we have

$$\mu \leq \min_{\substack{E_\mu v = 0 \\ v \in \mathfrak{D} \\ v \neq 0}} [(Hv, v)/(v, v)] \leq [(Hf, f)/(f, f)] = \mu. \tag{11}$$

This means that equality has to hold throughout (11) and that f must satisfy the Euler equation $Hf = \mu f$ (compare with Section 1.3). Since f is the eigenspace of μ, we get $E_\mu f = f$ while on the other hand, we already have $E_\mu f = 0$, which contradicts the assumption $f \neq 0$. Therefore, we must have $(Hf, f) > \mu(f, f)$, which contradicts (9) and proves the lemma.

Lemma 2. *Let H and A be two selfadjoint operators with domains $\mathfrak{D}(H)$ and $\mathfrak{D}(A)$ and let $E_\lambda(H)$, $E_\lambda(A)$, $\mathfrak{M}_\lambda(H)$, and $\mathfrak{M}_\lambda(A)$ be the corresponding spectral projections and subspaces. If $\mathfrak{D}(H) \subset \mathfrak{D}(A)$ and $(Af, f) \leq (Hf, f)$ for every $f \in \mathfrak{D}(H)$, then*

$$\dim \mathfrak{M}_\lambda(H) \leq \dim \mathfrak{M}_\lambda(A) \tag{12}$$

for every λ.

PROOF. Suppose that there is a point μ such that

$$\dim \mathfrak{M}_\mu(A) < \dim \mathfrak{M}_\mu(H).$$

Then, there exists a vector $f \in \mathfrak{D}(H)$, $(f, f) = 1$, such that $f \in \mathfrak{M}_\mu(H)$ and $f \perp \mathfrak{M}_\mu(A)$ (compare with Section 3.1). This means that $E_\mu(H)f = f$, while $E_\mu(A)f = 0$. Using the same analysis as given above, $E_\mu(A)f = 0$ implies the strict inequality

$$\mu < (Af, f).$$

On the other hand, $E_\mu(H)f = f$ implies $[I - E_\mu(H)]f = 0$, so that, from the spectral formula, we have

$$(Hf, f) = \int_{-\infty}^{\infty} \lambda \, d(E_\lambda f, f) = \int_{-\infty}^{\mu+0} \lambda \, d(E_\lambda f, f) \le \mu.$$

Therefore, we have

$$\mu < (Af, f) \le (Hf, f) \le \mu,$$

which is a contradiction and proves the lemma.

For operators of class $\mathscr{S}$, this lemma shows that under a given number μ a bigger operator cannot have more eigenvalues than a smaller operator.

Kato [K5] characterizes the eigenvalues by

$$\lambda_k = \inf\{\lambda \mid \dim \mathfrak{M}_\lambda \ge k\}. \tag{13}$$

This is actually a special case of the minimum–maximum principle (2.2.3). In fact, for $\lambda_k \le \lambda < \lambda_\infty$, we obtain $\mathfrak{M}_\lambda = \mathfrak{U}_j$ for some $j \le k$. By Poincaré's inequality (2.2.1), we have

$$\lambda_k \le \max_{u \in \mathfrak{M}_\lambda} R(u) = \lambda_j,$$

while

$$\lambda_k = \max_{u \in \mathfrak{M}_{\lambda k}} R(u).$$

Therefore, we have

$$\lambda_k = \min_{\substack{\mathfrak{M}_\lambda \\ \dim \mathfrak{M}_\lambda \ge k}} \max_{u \in \mathfrak{M}_\lambda} R(u).$$

The characterization (13) and Lemma 2 lead to another proof of the monotonicity principle (2.5.3). Actually, this proof was given after the original proof of Weyl's but before the proof of Section 2.5.

In view of Lemma 2, we have

$$\{\lambda \mid \dim \mathfrak{M}_\lambda(H) \ge k\} \subset \{\lambda \mid \dim \mathfrak{M}_\lambda(A) \ge k\}.$$

Therefore,

$$\begin{aligned} \lambda_k(A) &= \inf\{\lambda \mid \dim \mathfrak{M}_\lambda(A) \ge k\} \\ &\le \inf\{\lambda \mid \dim \mathfrak{M}_\lambda(H) \ge k\} = \lambda_k(H). \end{aligned}$$

Of course, all proofs are connected with the original proof of Weyl's. For instance, Weyl's test function u_0 appears implicitly in $E_\mu(H)f=f$, while the orthogonality conditions in (3.1.1) appear implicitly in $E_\mu(A)f=0$.

Returning to the discussion of helium, in order to show the existence of points μ such that $\dim \mathfrak{M}_\mu < \infty$, let us compare the helium wave operator with the wave operator

$$A = -\tfrac{1}{2}\,\Delta_1 u - \tfrac{1}{2}\,\Delta_2\, u - (2/r_1)u - (2/r_2)u.$$

This operator A is the Hamiltonian of a system composed of two independent hydrogenlike atoms and admits a unique selfadjoint extension, say $\tilde{A}$, having the same domain as $\tilde{H}$. The eigenvalues of $\tilde{A}$ are well known and given by

$$-2[(1/{n_1}^2) + (1/{n_2}^2)] \qquad (n_1, n_2 = 1, 2, \ldots)$$

with multiplicities ${n_1}^2{n_2}^2$, and have explicitly known eigenfunctions (see Kemble [20]). Since the essential spectrum of A consists of the interval $[-2, \infty)$, the lower part of the spectrum begins with isolated eigenvalues. For all $\mu < -2$ we have therefore

$$\dim \mathfrak{M}_\mu(\tilde{A}) < \infty.$$

As we have

$$\tilde{H} = \tilde{A} + B,$$

where B is the multiplicative (symmetric) operator, $Bu = (1/r_{12})u$, and $(Bu, u) \geq 0$, we have by Lemma 2

$$\dim \mathfrak{M}_\mu(\tilde{H}) \leq \dim \mathfrak{M}_\mu(\tilde{A}) < \infty$$

for all $\mu < -2$.

In order to show the existence of points $\mu < -2$ such that $0 < \dim \mathfrak{M}_\mu(\tilde{H})$, Kato explicitly determines points μ and corresponding finite-dimensional subspaces $\mathfrak{V}$, satisfying, as in Lemma 1, the inequalities

$$(\tilde{H}f, f) \leq \mu(f, f) \qquad (f \in \mathfrak{V}).$$

Let us also add that Zislin and Sigalov have proved, among other results, that there is a sequence of eigenvalues at the beginning of the spectrum of $\tilde{H}$ whose eigenfunctions depend only upon r_1, r_2, and r_{12} and are symmetric in the spatial coordinates of the two electrons (see [Z4, ZS1, ZS2]).

For further details, refer to [K2–3, Z4, ZS1–2].

8. Application of the Special Choice to the Helium Atom

The first and most outstanding application of the special choice was Bazley's determination of lower bounds for the two lowest eigenvalues (corresponding to eigenfunctions possessing the symmetry mentioned at the end of the previous section) of the Schrödinger equation for helium [ZS1–2]. We give here a brief sketch of his work.

For more details, refer to [K2–3, B1–3, BF1].

Following Bazley [B1], we put $H = A + B$ (see also Section 7), where, for simplicity, we now write A for $\tilde{A}$ and H for $\tilde{H}$, and B is the nonnegative operator given by

$$Bu = (1/r_{12})u.$$

The spectrum of A begins with isolated eigenvalues

$$\lambda_k^{(0)} = -2[1 + (1/k^2)] \qquad (k = 1, 2, \ldots)$$

having eigenfunctions

$$u_1^{(0)} = -2(1/4\pi)R_{10}(r_1)R_{10}(r_2),$$

$$u_k^{(0)} = (1/\sqrt{24}\pi)[R_{10}(r_1)R_{k0}(r_2) + R_{10}(r_2)R_{k0}(r_1)] \qquad (k = 2, 3, \ldots),$$

where the elements R_{k0} are the normalized hydrogen radial wave functions.

By observing that B is easily invertible and yields the special choice $p_i = r_{12}u_i^{(0)}$ $(i = 1, 2, 3)$, Bazley solved the third intermediate problem and obtained the roots -3.063_7, -2.165_5, and -2.039_2. The eigenvalues $\lambda_4^{(0)}, \lambda_5^{(0)}, \ldots$ are persistent in view of the special choice (see Section 4). Since

$$\lambda_7^{(0)} < -2.039_2 < \lambda_8^{(0)},$$

the eigenvalues of the third intermediate problem are given by

$$\begin{aligned}
\lambda_1^{(3)} &= -3.063_7\\
\lambda_2^{(3)} &= -2.165_5\\
\lambda_3^{(3)} &= \lambda_4^{(0)} = -2.125\\
\lambda_4^{(3)} &= \lambda_5^{(0)} = -2.080\\
\lambda_5^{(3)} &= \lambda_6^{(0)} = -2.055_6\\
\lambda_6^{(3)} &= \lambda_7^{(0)} = -2.040_8\\
\lambda_7^{(3)} &= -2.039_2\\
\lambda_8^{(3)} &= \lambda_8^{(0)}\\
\lambda_9^{(3)} &= \lambda_9^{(0)}\\
&\ldots.
\end{aligned}$$

Using the Rayleigh–Ritz upper bounds due to Kinoshita [K6] and the above *lower* bounds, Bazley gave the inequalities

$$-3.063_7 \leq \lambda_1 \leq -2.9037237, \tag{1}$$

$$-2.165_5 \leq \lambda_2 \leq -2.1458. \tag{2}$$

9. Application of Intermediate Problems to Temple's Formula

A better lower bound for λ_1 of helium can be obtained by combining the numerical results of Section 8 with the inclusion theorem of Temple's [T1–2], which is the oldest of its kind.

Temple takes a "test function" w and obtains an explicitly computable interval which contains an eigenvalue.

Considering a problem for which λ_1 is a simple eigenvalue, we shall show that such an interval $[\alpha, \rho]$ can be chosen so that $\lambda_1 < \rho < \lambda_2$. Then, by definition, α will be a lower bound for λ_1.

Let $w \in \mathfrak{D}(H)$, $(w, w) = 1$ and let τ be any point that is in the resolvent set of H and below the essential spectrum. Then we have

$$\begin{aligned}(Hw - \tau w, Hw - \tau w) &= \sum_i |\lambda_i - \tau|^2 \, |(w, u_i)|^2 + \int_{\lambda_\infty - 0}^{\infty} |\lambda - \tau|^2 \, d(E_\lambda w, w) \\ &\geq \min_i |\lambda_i - \tau|^2 (w, w).\end{aligned}$$

Taking λ to be the λ_i for which the minimum occurs, we obtain the inequality

$$|\lambda - \tau| \leq [(Hw - \tau w, Hw - \tau w)]^{1/2} = \phi(\tau),$$

so that we have

$$\tau - \phi(\tau) \leq \lambda \leq \tau + \phi(\tau). \tag{1}$$

We now choose a fixed ρ $(\lambda_1 < \rho < \lambda_2)$ and compute a τ for which $\tau + \phi(\tau) = \rho$. Then, by (1), $\tau - \phi(\tau)$ is a lower bound for λ_1.

To obtain τ, we write

$$\begin{aligned}(\tau - \rho)^2 &= [\phi(\tau)]^2 \\ &= (Hw, Hw) - 2\tau(Hw, w) + \tau^2.\end{aligned}$$

Therefore, we have

$$\tau = [(Hw, Hw) - \rho^2]/[2(Hw, w) - 2\rho]$$

so that

$$\begin{aligned}\tau - \phi(\tau) &= \tau - \rho + \tau \\ &= 2\tau - \rho \\ &= [(Hw, Hw) - \rho^2 - \rho(Hw, w) + \rho^2]/[(Hw, w) - \rho] \\ &= [(Hw, Hw) - \rho(Hw, w)]/[(Hw, w) - \rho]\end{aligned}$$

which, by (1), yields *Temple's formula*

$$\lambda_1 \geq (Hw, w) - \{[(Hw, Hw) - (Hw, w)^2]/[\rho - (Hw, w)]\}. \tag{2}$$

Temple's formula is easy to apply. However, a meaningful numerical result depends first on the choice of the test function w, as in the Rayleigh–Ritz method (Section 2.3). Secondly, the knowledge of a number ρ ($\lambda_1 < \rho < \lambda_2$) implicitly requires the knowledge of a lower bound for λ_2 and an upper bound for λ_1. As late as 1959, Kinoshita [K6] used for the heilum atom the experimental value of λ_2 for ρ, which makes his result semiphenomenological. Such a procedure can be avoided because a close lower bound for λ_2 is provided by intermediate problems and can be taken as a rigorous value for ρ independent of spectroscopic observations. In this way, Temple's formula and intermediate problems enabled Bazley [B1] to obtain

$$-2.9037474 \leq \lambda_1 \leq -2.9037237,$$

which is an improvement of the lower bound for λ_1 given in (8.1).

Let us note in passing that Kinoshita used a 39-parameter trial function in applying the Rayleigh–Ritz method.

10. Application of Truncation to Quantum Theory

Bazley and Fox [BF1] applied the method of truncation to *the helium atom* and obtained exclusively by intermediate problems the inequality

$$-3.0008 \leq \lambda_1,$$

which is an improvement of the lower bound obtained by the special choice (8.1).

Another application is *the radial Schrödinger equation*, which we briefly recapitulate here. Bazley and Fox [BF1] consider the equation

$$-d^2\psi/dx^2 - z[(1 - e^{-\alpha x})/x]\psi = E\psi \tag{1}$$

on the interval $0 < x < \infty$ for α and z real and positive. The potential $-z(1 - e^{-\alpha x})/x$ behaves like the Coulomb potential $-z/x$ for large values of x, while near the origin it approaches $-\alpha z$. Furthermore, the potential differs from that of the hydrogenic wave equation by the positive term $ze^{-\alpha x}/x$.

In the treatment of (1), we are interested in the bound states only. While ordinarily one would consider E as the eigenvalue, here we fix the energy E and take the charge z as the eigenvalue. The numerical results of such calculations (done in sufficient detail) may then be inverted to give energy eigenvalues E as a function of charge. The advantage of taking z as the eigenvalue is that it eliminates the continuous spectrum.

We put $E = -k^2$ and introduce the transformations

$$t = 2kx, \qquad \varphi(t) = t^{-1/2}\psi(t), \tag{2}$$

so that (1) becomes

$$-\frac{d}{dt}\left(t\frac{d\varphi}{dt}\right) + \frac{t^2+1}{4t}\varphi = \lambda(1 - e^{-\alpha t/2k})\varphi, \tag{3}$$

where $\lambda = z/2k$. Equation (3) is an eigenvalue problem of the form

$$Au = \lambda(I - B)u, \tag{4}$$

where

$$Au = -\frac{d}{dt}\left(t\frac{du}{dt}\right) + \frac{t^2+1}{4t}u, \tag{5}$$

and

$$Bu = e^{-\alpha t/2k}u. \tag{6}$$

A suitable family of functions on which to define (3) consists of those functions that vanish at the origin, are square-integrable, and for which Au is square-integrable.

We note that $0 \le (u, Bu) \le 1$ for normalized u. Also, A has known eigenvalues and normalized eigenfunctions,

$$\lambda_i^{(0)} = i, \qquad (i = 1, 2, \ldots) \tag{7}$$

and

$$u_i^{(0)} = (t^{1/2}/i!i^{1/2})L_i'(t)e^{-t/2} \qquad (i = 1, 2, \ldots), \tag{8}$$

where L_i' is the first derivative of the ith Laguerre polynomial. Equation (3) has a pure point spectrum $\lambda_1 \le \lambda_2 \le \cdots$ diverging to infinity and satisfying

$$\lambda_i^{(0)} \le \lambda_i \qquad (i = 1, 2, \ldots).$$

We may proceed in direct analogy with the previous theory and introduce the eigenvalue problems

$$T_N u = \lambda(I - BP^k)u, \tag{9}$$

where T_N has been previously defined by (5.1) and P^k denotes a projection on arbitrary vectors $p_1, p_2, \ldots, p_k$ with respect to the inner product $[u, v] = (u, Bv)$. If we denote the ordered eigenvalues and eigenfunctions of (9) by $\lambda_i^{(N,k)}$ and $u_i^{(N,k)}$, respectively, we have

$$\lambda_i^{(N,k)} \leq \lambda_i \qquad (i = 1, 2, \ldots). \tag{10}$$

The $\lambda_i^{N,k}$ are monotonic in both N and k.

In the solution of (9), we choose

$$p_j = u_j^{(0)} \qquad (j = 1, 2, \ldots, k)$$

and find that

$$u^{(N,k)} = \lambda \sum_{i=1}^{k} \alpha_i \left\{ \sum_{j=1}^{N} \frac{(u_j^{(0)}, Bu_i^{(0)})u_j^{(0)}}{j - \lambda} + \frac{Bu_i^{(0)} - \sum_{j=1}^{N}(u_j^{(0)}, Bu_i^{(0)})u_j^{(0)}}{N + 1 - \lambda} \right\}.$$

Here, the values of α_i and λ are found as solutions of the algebraic system

$$0 = \sum_{i=1}^{k} \alpha_i \left\{ (u_j^{(0)}, Bu_i^{(0)}) + \lambda \sum_{s=1}^{N} \frac{(u_s^{(0)}, Bu_i^{(0)})(Bu_j^{(0)}, u_s^{(0)})}{s - \lambda} + \lambda \frac{(Bu_j^{(0)}, Bu_i^{(0)}) - \sum_{s=1}^{N}(u_s^{(0)}, Bu_i^{(0)})(Bu_j^{(0)}, u_s^{(0)})}{N + 1 - \lambda} \right\}. \tag{11}$$

As before, the multiplicity of each root is just the number of linearly independent solutions to the algebraic problem (11), and $\lambda_{N+1}^{(0)}$ appears as an eigenvalue of infinite multiplicity.

For this example, Bazley and Fox fixed the value of α and chose $k = \alpha/2$ so that $E = -\alpha^2/4$. They computed the lower bounds from (11) for several values of N and k.

Upper bounds obtained by solving a fourth-order Rayleigh–Ritz problem based on the trial functions $u_1^{(0)}$, $u_2^{(0)}$, $u_3^{(0)}$, and $u_4^{(0)}$ together with the lower bounds $\lambda_i^{(3,2)}$ provided the following estimates:

$$1.2587\alpha \leq z_1 \leq 1.2590\alpha,$$

$$2.3944\alpha \leq z_2 \leq 2.4164\alpha,$$

$$3.4207\alpha \leq z_3 \leq 3.5576\alpha.$$

Chapter Six

Various Other Methods and Their Connections with Intermediate Problems

1. Quadratic Forms and Intermediate Problems

In this chapter, we shall discuss some methods for the determination of lower bounds for eigenvalues that are outgrowths of the methods of intermediate problems.

There are important operators, especially differential operators, for which the decomposition $A + B$ used in the previous chapter is not obvious. However, Bazley and Fox [BF5] noticed that it is sometimes possible to decompose the numerator $J(u) = (Tu, u)$ of the Rayleigh quotient

$$R(u) = (Tu, u)/(u, u)$$

as

$$J(u) = J^0(u) + J'(u),$$

where the operator $A^{(0)}$ corresponding to $J^0(u)$ has a known spectral resolution and $J'(u)$ is a positive-semidefinite form. For a significant application, see Section 2. Moreover, it turns out that there exists an operator C having $\mathfrak{D}(C) \subset \mathfrak{H}$ and $\mathfrak{R}(C)$ contained in another Hilbert space $\mathfrak{H}'$ such that

$$J'(u) = (Cu, Cu)' \qquad [u \in \mathfrak{D}(C)].$$

Since we have $\mathfrak{D}(T)$ dense in $\mathfrak{H}$ and

$$\mathfrak{D}(T) \subset \mathfrak{D}(J) = \mathfrak{D}(J^0) \cap \mathfrak{D}(J') \subset \mathfrak{D}(J') = \mathfrak{D}(C),$$

it follows that C is densely defined and therefore has a uniquely defined adjoint operator C^*, where

$$(Cu, v)' = (u, C^*v)$$

for $u \in \mathfrak{D}(C)$ and $v \in \mathfrak{D}(C^*) \subset \mathfrak{H}'$. We now choose a finite number of vectors $p_1, p_2, \ldots, p_n \in \mathfrak{D}(C^*)$ and let P^n denote the projection operator (orthogonal in $\mathfrak{H}'$) onto $\text{sp}\{p_1. p_2, \ldots, p_n\}$. Then we obtain the intermediate quadratic form

$$\begin{aligned} J^n(u) &= J^0(u) + (P^n Cu, Cu), \\ &= (A^{(0)}u, u) + (C^*P^nCu, u), \end{aligned} \tag{1}$$

which, by Bessel's inequality, satisfies the inequalities

$$J^0(u) \leq J^n(u) \leq J(u). \tag{2}$$

The operator corresponding to $J^n(u)$ is given by

$$A^{(n)} = A^{(0)} + C^*P^nC. \tag{3}$$

By the *monotonicity principle* and inequalities (2) its eigenvalues $\{\lambda_j^{(n)}\}$ satisfy

$$\lambda_i^{(0)} \leq \lambda_i^{(n)} \leq \lambda_i^{(\infty)} \qquad (i = 1, 2, \ldots).$$

Here, the superscripts 0 and ∞ are again used to denote the eigenvalues of the *base* and *given* problems, respectively.

In order to solve the eigenvalue problem $A^{(n)}u = \lambda u$, we need to solve

$$A^{(0)}u + \sum_{i,\, j=1}^{n} (u, C^*p_i)\beta_{ij}\, C^*p_j = \lambda u, \tag{4}$$

where $\{\beta_{ij}\}$ is the matrix inverse to $\{(p_i, p_j)'\}$. To derive this equation, we proceed exactly as we have already done in Chapter 5, Section 2. For instance, let us consider here the case $n = 1$. Then,

$$\begin{aligned} C^*P^1Cu &= C^*[(p_1, Cu)']p_1 \\ &= (C^*p_1, u)C^*p_1. \end{aligned}$$

The solution of (4) now can be obtained by the same rules as in Chapter 5.

Sometimes, the unknown operator already has the form $A^{(0)} + B = A^{(0)} + C^*C$. Since the quadratic form here is

$$\begin{aligned} J(u) &= (A^{(0)}u, u) + (C^*Cu, u) \\ &= (A^{(0)}u, u) + (Cu, Cu)', \end{aligned}$$

we could apply either the methods of Chapter 5 or that of the present section. In applications, we often have in the latter case the advantage that the elements $\{p_i\}$ may be chosen from a wider class of functions, namely, as mentioned above, from $\mathfrak{D}(C^*)$ rather than $\mathfrak{D}(B)$. Usually, this will mean that the vectors $\{p_i\}$ will have to satisfy fewer boundary conditions and fewer differentiability conditions.

EXAMPLE. Let us consider the ordinary differential eigenvalue problem stated by

$$(d^2/dx^2)\,(1 + a\cos^2 x)(d^2u/dx^2) - \lambda u = 0 \qquad (-\pi/2 < x < \pi/2),$$

$$u(\pi/2) = u''(\pi/2) = 0, \qquad u(-x) = u(x) \qquad (-\pi/2 < x < \pi/2),$$

in which a is a nonnegative real constant. In the notation used above, the spaces, operators, and quadratic forms are given by

$$\mathfrak{H} = \mathfrak{H}' = \{u \in \mathscr{L}^2(-\pi/2, \pi/2) \mid u(-x) = u(x)\},$$

$$Tu = (d^2/dx^2)(1 + a\cos^2 x)\, d^2u/dx^2, \qquad u(\pi/2) = u''(\pi/2) = 0,$$

$$Au = A^{(0)}u = d^4u/dx^4, \qquad u(\pi/2) = u''(\pi/2) = 0,$$

$$Bu = a(d^2/dx^2)(\cos^2 x)(d^2u/dx^2), \qquad u(\pi/2) = 0, \quad u''\cos^2 x\,|_{\pi/2} = 0,$$

$$J(u) = \int_{-\pi/2}^{\pi/2} (1 + a\cos^2 x)\,|d^2u/dx^2|^2\, dx, \qquad u(\pi/2) = 0,$$

$$J^0(u) = \int_{-\pi/2}^{\pi/2} |d^2u/dx^2|^2\, dx, \qquad u(\pi/2) = 0,$$

$$J'(u) = a\int_{-\pi/2}^{\pi/2} \cos^2 x\,|d^2u/dx^2|^2\, dx, \qquad u(\pi/2) = 0,$$

$$Cu = -a^{1/2}(\cos x)\, d^2u/dx^2, \qquad u(\pi/2) = 0,$$

$$C^*v = -a^{1/2}\,(d^2/dx^2)(v\cos x), \qquad v\cos x\,|_{\pi/2} = 0.$$

From the above, we see that C^* only requires one boundary condition and two derivatives, while B requires two boundary conditions and four derivatives.

Let us note that this variant of the method of intermediate problems is sometimes called the B^*B method. In our notation, we had to replace B^*B by C^*C in order not to contradict our previous notations.

2. Application of Quadratic Forms to Free and Cantilever Plates

As mentioned in Section 2.3, the original problem of Ritz's was the determination of upper bounds for the eigenvalues of a vibrating, free, square plate. A lower bound for the first eigenvalue was given by Kato, Fujita, Nakata, and Newman [KFNN1], who used an inclusion theorem similar to that of Section 5.9.

Bazley, Fox, and Stadter [BFS1] applied the method of Section 1 to obtain lower bounds for a great number of eigenvalues for this famous problem. Still more important and more difficult is the vibration of a rectangular plate clamped on one side and free on the others, a problem treated for the first time by Bazley, Fox, and Stadter [BFS1]. Such a plate may be thought of as the wing of an airplane. Since the procedures in both cases are closely related, we limit ourselves to the case of the cantilever plate. The vibration is described by the classical differential eigenvalue problem

$$\Delta\,\Delta u - \lambda u = 0 \qquad \text{in} \quad \Omega$$

with the conditions

$$u = \frac{\partial u}{\partial n} = 0 \qquad \text{on} \quad \Gamma_1,$$

$$\left.\begin{aligned} \sigma\,\Delta u + (1-\sigma)\frac{\partial^2 u}{\partial n^2} &= 0 \\ \frac{\partial(\Delta u)}{\partial n} + (1-\sigma)\frac{d}{ds}\left[\frac{\partial^2 u}{\partial n\,\partial t}\right] &= 0 \end{aligned}\right\} \qquad \text{on} \quad \Gamma_2,$$

and $\partial^2 u/\partial n\,\partial t$ continuous in s at the free corners. Here, Ω is the rectangle

$$-a/2 \le x \le a/2, \qquad -b/2 \le y \le b/2, \tag{1}$$

Γ_1 is the clamped side $y = -b/2$, σ is Poisson's ratio, Γ_2 is the free portion of the boundary $x = \pm a/2$ and $y = b/2$, $\partial/\partial n$ and $\partial/\partial t$ are the derivatives in the exterior normal and tangential directions on the boundary, and d/ds is the derivative with respect to arc length along the boundary.

The corresponding variational problem is

$$\min J(u) = \min \iint_\Omega \{\sigma(\Delta u)^2 + (1-\sigma)[u_{xx}^2 + u_{yy}^2 + 2u_{xy}^2]\}\,dx\,dy \tag{2}$$

defined for sufficiently smooth functions u satisfying $(u, u) = 1$ and the boundary conditions

$$u = \partial u/\partial n = 0 \qquad \text{on} \quad \Gamma_1.$$

The conditions on Γ_1 are prescribed, while the conditions on Γ_2 and at the free corners are called *natural boundary conditions* because they are automatically satisfied by the minimizing functions (see Section 4.12).

The quadratic form J given by Eq. (2) can be decomposed for our purposes as

$$J = J^0 + J', \tag{3}$$

in which J^0 is given by

$$J^0(u) = (1 - \sigma) \iint_\Omega (u_{xx}^2 + u_{yy}^2)\, dx\, dy \qquad (u = \partial u/\partial n = 0 \quad \text{on} \quad \Gamma_1)$$

and J' by

$$J'(u) = \iint_\Omega [\sigma(\Delta u)^2 + 2(1 - \sigma)u_{xy}^2]\, dx\, dy \qquad (u = \partial u/\partial n = 0 \quad \text{on} \quad \Gamma_1).$$

The variational problem

$$\delta\left[J^0(u) \Big/ \iint_\Omega u^2\, dx\, dy\right] = 0$$

gives rise to the corresponding differential eigenvalue problem

$$\begin{aligned}(1 - \sigma)(u_{xxxx}^{(0)} + u_{yyyy}^{(0)}) - \lambda^{(0)}u^{(0)} = 0 \qquad &\text{in} \quad \Omega,\\ u^{(0)} = \partial u^{(0)}/\partial n = 0 \qquad &\text{on} \quad \Gamma_1,\\ \partial^2 u^{(0)}/\partial n^2 = \partial^3 u^{(0)}/\partial n^3 = 0 \qquad &\text{on} \quad \Gamma_2.\end{aligned} \tag{4}$$

The conditions on Γ_2 are again natural boundary conditions.

This problem can be used as a base problem, since it can be solved explicitly by separation of variables. In this way, we obtain

$$u_{r,s}^{(0)}(x, y) = [2/(ab)^{1/2}]d_r(2x/a)e_s(2y/b)$$

and

$$\lambda_{r,s}^{(0)} = (1 - \sigma)[(2/a)^4 \alpha_r{}^4 + (2/b)^4 \beta_s{}^4],$$

where d_r and α_r satisfy

$$d_r'''' - \alpha_r{}^4\, d_r = 0, \qquad d_r''(-1) = d_r'''(-1) = d_r''(1) = d_r'''(1) = 0, \tag{5}$$

and e_s and β_s satisfy

$$\underline{\ ^s}\, e_s'''' \quad \beta e_s = 0, \qquad e_s(-1) = e_s'(-1) = e_s''(1) = e'''(1) = 0. \tag{6}$$

Incidentally, (5) is the problem of a free beam and (6) is the problem of a clamped-free beam.

Let us now consider the semipositive form $J'(u)$. In order to write this form as $(Cu, Cu)'$, we introduce a second Hilbert space $\mathfrak{H}' = \mathfrak{H} \times \mathfrak{H}$, each element of which is a pair of vectors $\{u^1, u^2\}$ $(u^1, u^2 \in \mathfrak{H})$. Addition and scalar multiplication are defined in the obvious manner and the inner product is defined by

$$(\{u^1, u^2\}, \{v^1, v^2\})', = (u^1, v^1) + (u^2, v^2),$$

where, as usual,

$$(u, v) = \iint_\Omega uv \, dx \, dy.$$

The operator C is defined by the equation

$$Cu = \{\sigma^{1/2} \Delta u, 2^{1/2}(1 - \sigma)^{1/2} u_{xy}\}$$

for sufficiently smooth functions u satisfying

$$u = \partial u/\partial n = 0 \qquad \text{on} \quad \Gamma_1. \tag{7}$$

Then, we have the desired equality

$$J'(u) = (Cu, Cu)'.$$

The adjoint C^* is obtained by integration by parts of the integral

$$(Cu, p)' = \iint_\Omega [\sigma^{1/2}(\Delta u)p^1 + 2^{1/2}(1 - \sigma)^{1/2} u_{xy} p^2] \, dx \, dy,$$

where $p = \{p^1, p^2\}$ and u satisfies (7). Then, the adjoint C^* is given by

$$C^*\{p^1, p^2\} = \sigma^{1/2} \Delta p^1 + 2^{1/2}(1 - \sigma)^{1/2} p^2_{xy}$$

where p^1 and p^2 are sufficiently smooth functions in Ω satisfying the boundary conditions

$$\begin{aligned} &p^1 = 0, \qquad \sigma^{1/2} p_x{}^1 + 2^{1/2}(1 - \sigma)^{1/2} p_y{}^2 = 0 \qquad \text{on} \quad x = \pm a/2, \\ &p^1 = 0, \qquad \sigma^{1/2} p_y{}^1 + 2^{1/2}(1 - \sigma)^{1/2} p_x{}^2 = 0 \qquad \text{on} \quad y = b/2, \\ &p^2(\pm a/2, b/2) = 0. \end{aligned} \tag{8}$$

For the numerical computations, it is important to note that all these conditions (8) are automatically satisfied by functions p^1, p^2 that satisfy the simpler boundary conditions

$$p^1 = \partial p^1/\partial n = 0, \qquad p^2 = 0 \quad \text{on} \quad \Gamma_2.$$

Now, the intermediate problems can be constructed as in Section 1 and together with the Rayleigh–Ritz method yield excellent numerical results for numerous eigenvalues and ratios b/a. Four numerical tables, each containing ten eigenvalues, are given in [BFS2] (see Table B5).

3. The Rhombical Membrane

Among the many variational problems to be studied are those in which domains of various shapes occur. Intermediate problems can also be applied to questions of this type. In particular, we give here a method for obtaining upper and lower bounds for the eigenvalues of a rhombical membrane fixed on its edges. Bounds for the *first* eigenvalue have been obtained by Polya and Szegö [24], Weinberger [W5], and Hooker and Protter [HP1]. More recently, Stadter [S2], using intermediate problems, obtained upper and lower bounds for the first *ten* eigenvalues, corresponding to eigenfunctions which are even in both variables. Let us briefly describe Stadter's procedure.

We seek upper and lower bounds to the eigenvalues of vibrating uniform rhombical membranes with sides of unit length and skew angle

$$\theta \quad (0 \leqq \theta < \pi/2),$$

as shown in Fig. 1. We consider two cases: the membrane fixed on all edges, and the membrane fixed on two opposite edges, free on the other edges (Fig. 1).

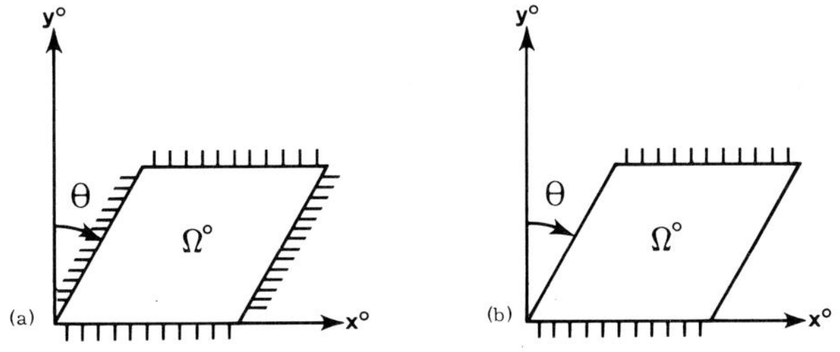

FIG. 1. (a) Fixed membrane. (b) Fixed-free membrane. [Reprinted with permission from J. T. Stadter, *SIAM J. Appl. Math.* **14**, 324–341 (1966). Copyright 1966 by Soc. Ind. Appl. Math. All rights reserved.]

The membrane covers the rhombical domain Ω^0 with boundary Γ^0. We designate the fixed portion of the boundary by $\Gamma_1{}^0$ and the free portion by $\Gamma_2{}^0$. The eigenvalue problem is stated by

$$\begin{aligned} -\Delta u^0 - \lambda u^0 &= 0 \quad \text{in } \Omega^0, \\ u^0 &= 0 \quad \text{on } \Gamma_1{}^0 \\ \partial u^0/\partial n &= 0 \quad \text{on } \Gamma_2{}^0. \end{aligned} \tag{1}$$

We introduce the Hilbert space $\mathfrak{H}^0$ equal to $\mathscr{L}^2(\Omega^0)$, the space of real-valued functions square-integrable over Ω^0. The inner product in $\mathfrak{H}^0$ is given by

$$(u, v)^0 = \iint_{\Omega} u(x^0, y^0)v(x^0, y^0)\, dx^0\, dy^0. \tag{2}$$

In the following, we will state only restrictions of the domains of operators and quadratic forms. These restricted domains will be sufficiently large that the operators on them will be essentially selfadjoint or closeable and that the quadratic forms will be closeable. In this way, we avoid the complications involved in a full description of these domains. In each case, the functions employed in the constructions will belong to the restricted domains.

We may consider (1) as the eigenvalue problem for a selfadjoint operator A^0 having domain

$$\mathfrak{D}(A^0) = \{u^0 \in \mathscr{C}^2(\Omega^0) \cap \mathscr{C}^1(\overline{\Omega}^0) \,|\, u^0 = 0 \quad \text{on} \quad \Gamma_1{}^0, \quad \partial u^0/\partial n = 0 \quad \text{on} \quad \Gamma_2{}^0\}$$

in $\mathfrak{H}^0$ and associated quadratic form J_{A^0} given by

$$J_{A^0}(u^0) = \iint_{\Omega^0} [(u^0_{x^0})^2 + (u^0_{y^0})^2]\, dx^0\, dy^0, \tag{3}$$

with domain $\mathfrak{D}(J_{A^0}) = \{u^0 \in \mathscr{C}^1(\Omega^0) \,|\, u^0 = 0 \text{ on } \Gamma_1{}^0\}$.

We introduce the affine coordinate transformation $\alpha[x, y] = [x^0, y^0]$, given by

$$x^0 = x + y \sin\theta, \quad y^0 = y \cos\theta, \tag{4}$$

which maps the unit square Ω with boundary Γ onto the rhombus Ω^0; Γ_1 corresponds to $\Gamma_1{}^0$ and Γ_2 corresponds to $\Gamma_2{}^0$.

The transformation α has Jacobian ρ equal to $\cos\theta$ and induces an isometric isomorphism V of $\mathfrak{H}^0$ onto $\mathfrak{H}$, where $\mathfrak{H}$ is $\mathscr{L}^2(\Omega)$, the space of real-valued functions square-integrable over Ω. The space $\mathfrak{H}$ has inner product

$$(u, v) = \iint_{\Omega} u(x, y)v(x, y)\, dx\, (dy. \tag{5}$$

Since

$$\iint_{\Omega^0} u^0(x^0, y^0)v^0(x^0, y^0)\, dx^0\, dy^0 = \iint_{\Omega} u^0(\alpha[x, y])v^0(\alpha[x, y])\rho\, dx\, dy, \tag{6}$$

the isomorphism V is given by

$$Vu^0(x^0, y^0) = \rho^{1/2}u^0(\alpha[x, y]). \tag{7}$$

The operator A in $\mathfrak{H}$, isomorphic to A^0, is given by

$$A = VA^0V^{-1},$$

with domain $\mathfrak{D} = V\mathfrak{D}(A^0)$. The quadratic form J associated with A is given by

$$J(u) = \iint_\Omega (\sec^2 \theta)\,[u_x{}^2 - 2(\sin \theta)u_x u_y + u_y{}^2]\,dx\,dy, \tag{8}$$

with domain

$$\mathfrak{D}(J) = V\mathfrak{D}(J_{A^0}) = \{u \in \mathscr{C}^1(\overline{\Omega})\,|\,u = 0 \quad \text{on} \quad \Gamma_1\}.$$

Since A is isomorphic to A^0, A and A^0 have the same eigenvalues. Thus, bounds for the eigenvalues of A are also bounds for those of A^0, so that it sufficies to find bounds for the eigenvalues of A.

The quadratic form (8) admits two decompositions, namely

$$\begin{aligned} J(u) &= (\sec^2 \theta)(1 + \sin \theta) \iint_\Omega (u_x{}^2 + u_y{}^2)\,dx\,dy \\ &\quad -(\sec^2 \theta)(\sin \theta) \iint_\Omega (u_x + u_y)^2\,dx\,dy \\ &= J_{A^+}(u) - J^+(u), \end{aligned} \tag{9}$$

and

$$\begin{aligned} J(u) &= (\sec^2 \theta)(1 - \sin \theta) \iint_\Omega (u_x{}^2 + u_y{}^2)\,dx\,dy \\ &\quad + (\sec^2 \theta)(\sin \theta) \iint_\Omega (u_x - u_y)^2\,dx\,dy \\ &= J_{A^-}(u) + J^-(u). \end{aligned} \tag{10}$$

The quadratic forms J^+ and J^- can be written as

$$J^+(u) = (B^+u, B^+u) \qquad \text{and} \qquad J^-(u) = (B^-u, B^-u),$$

where

$$B^+u = (\sec^2 \theta \sin \theta)^{1/2}(\partial/\partial x + \partial/\partial y)u$$

and

$$B^-u = (\sec^2 \theta \sin \theta)^{1/2}(\partial/\partial x - \partial/\partial y)u.$$

Moreover, the eigenvalue problems corresponding to J_{A^+} and J_{A^-} are explicitly solvable by separation of variables since each quadratic form is just a (constant) multiple of the quadratic form for the vibrating square membrane Ω. The adjoints of B^+ and B^-, obtained through integration by parts, are

$$(B^+)^*u = -(\sec^2\theta \sin\theta)^{1/2}(\partial/\partial x + \partial/\partial y)u$$

for u vanishing on Γ_2, having continuous first derivatives in the direction [1, 1], and being piecewise continuous in the direction [1, −1], and

$$(B^-)^*u = -(\sec^2\theta \sin\theta)^{1/2}(\partial/\partial x - \partial/\partial y)u$$

for u vanishing on Γ_2, having continuous first derivatives in the direction [1, −1], and being piecewise continuous in the direction [1, 1], so that the method of Section 1 can be applied.

For numerical results, see Tables B.9–B.10.

4. Bazley–Fox Method for Sums of Solvable Operators

In the intermediate problems of the second type, the unknown operator T appeared as the sum $A + B$, where A has known eigenelements and B is a positive operator. However, an important variant of this theory was discovered by Bazley and Fox, namely when the unknown operator T can be written as

$$T = A^{[1]} + A^{[2]}, \qquad \mathfrak{D}(T) = \mathfrak{D}(A^{[1]}) \cap \mathfrak{D}(A^{[2]}),$$

where each of the $\mathscr{S}$-operators $A^{[1]}$ and $A^{[2]}$ has a known spectral resolution and neither is required to be positive. (The procedure due to Bazley and Fox is of course valid for a sum of more than two operators.) While one could use the second type of intermediate problems by writing

$$T = A^{[1]} - cI + A^{[2]} + cI,$$

where $c > 0$ is sufficiently large so that $B = A^{[2]} + cI$ is positive, Bazley and Fox use exclusively truncation in the following way.

We let $\{\lambda_i^{[j]}\}$ and $\{u_i^{[j]}\}$ $(i = 1, 2, \ldots; j = 1, 2)$ denote the eigenvalues and eigenvectors of $A^{[1]}$ and $A^{[2]}$. For fixed nonnegative integers n_1 and n_2, we let P_{n_j} be the orthogonal projection operator onto the subspace $\mathfrak{U}_{n_j}^{[j]} = \mathrm{sp}\{u_1^{[j]}, u_2^{[j]} \ldots, u_{n_j}^{[j]}\}$ $(j = 1, 2)$. Letting $\mathbf{n}$ denote the pair (n_1, n_2), we introduce a new operator $T^{\mathbf{n}}$ which is defined by taking the sum of the *truncations* of $A^{[1]}$ and $A^{[2]}$, namely

$$T^{\mathbf{n}}u = \sum_{j=1}^{2} \left[\sum_{i=1}^{n_j} \lambda_i^{[j]}(u, u_i^{[j]})u_i^{[j]} + \lambda_{n_j+1}^{[j]}(I - P_{n_j})u \right], \tag{1}$$

where n_1 and n_2 are chosen so that $\lambda_{n_1}^{[1]} < \lambda_{n_1+1}^{[1]}$ and $\lambda_{n_2}^{[2]} < \lambda_{n_2+1}^{[2]}$. Since $(T^{\mathbf{n}}u, u) \le (Tu, u)$ and $(T^{\mathbf{n}}u, u)$ is nondecreasing for increasing n_1 and n_2, the eigenvalues $\lambda_1{}^{\mathbf{n}} \le \lambda_2{}^{\mathbf{n}} \le \cdots$ of $T^{\mathbf{n}}$ form a nondecreasing sequence of lower bounds for those of T. Let us note here that the improvement of lower bounds is not achieved by solving intermediate problems relative to a fixed truncated base operator (see Section 4.10), but by increasing the indices of truncation.

Let $\mathfrak{B}_{\mathbf{n}} = \mathfrak{U}_{n_1}^{[1]} + \mathfrak{U}_{n_2}^{[2]}$ (not necessarily a direct sum). Then, from (1), it is clear that every vector v in $\mathfrak{B}_{\mathbf{n}}{}^{\perp}$ is an eigenvector of $T^{\mathbf{n}}$ having eigenvalues $\lambda_{n_1+1}^{[1]} + \lambda_{n_2+1}^{[2]}$. In fact, since $(v, u_i^{[j]}) = 0\ (i = 1, 2, \ldots, n_j\,; j = 1, 2)$, we have $[I - P_{n_j}]v = v$, and therefore

$$T^{\mathbf{n}}v = [\lambda_{n_1+1}^{[1]} + \lambda_{n_2+1}^{[2]}]v.$$

This eigenvalue $\lambda_{n_1+1}^{[1]} + \lambda_{n_2+1}^{[2]}$, plays the part of the eigenvalue $\lambda_{n+1}^{(0)}$ of infinite multiplicity in ordinary truncation theory (Section 4.10). We have thus decomposed $T^{\mathbf{n}}$ into two parts relative to two mutually orthogonal subspaces in order to apply Theorem A.4. Therefore, the remaining eigenvalues of $T^{\mathbf{n}}$ are the eigenvalues of the part of $T^{\mathbf{n}}$ in the finite-dimensional space $\mathfrak{B}_{\mathbf{n}}$ and are explicitly computable.

In ordinary truncation, the whole spectrum is known by inspection. Here, however, we had to decompose the operator into two parts and determine the spectrum on a finite-dimensional space.

One of the significant numerical applications of the above method is the computation of lower bounds for the eigenvalues of a clamped plate, a problem which had already been treated by the first type of intermediate problems (see Section 4.12).

We recall that the eigenvalue problem is

$$\begin{aligned} \Delta^2 u - \lambda u &= 0 \qquad \text{in} \quad \Omega, \\ u = \partial u/\partial n &= 0 \qquad \text{on} \quad \Gamma, \end{aligned}$$

where Ω is again a rectangle and Γ is its boundary. The operator

$$\Delta^2 = \partial^4/\partial x^4 + 2(\partial^4/\partial x^2\,\partial y^2) + \partial^4/\partial y^4$$

is decomposed into $A^{[1]} + A^{[2]}$, where

$$\begin{aligned} A^{[1]}u &= u_{xxxx} + u_{yyyy} \qquad \text{in} \quad \Omega, \\ u &= \partial u/\partial n = 0 \qquad \text{on} \quad \Gamma, \end{aligned} \tag{2}$$

and

$$\begin{aligned} A^{[2]}u &= 2u_{xxyy} \qquad \text{in} \quad \Omega, \\ u &= 0 \qquad \text{on} \quad \Gamma. \end{aligned} \tag{3}$$

Both eigenvalue problems can be readily solved by separation of variables (see [AD1]). Using the explicit solutions of (2) and (3), Bazley, Fox, and Stadter [BFS2] applied the truncation method described above to obtain lower bounds for fifteen eigenvalues for various side-length ratios (see Table B.3).

5. The Trefftz–Fichera Method for Integral Operators

In this section, we discuss a method of obtaining lower bounds for the eigenvalues of integral operators. While this method in itself has no direct connection with intermediate problems, we need the following exposition as a preparation for the next section. The method was introduced in 1933 by Trefftz [T7]† and was nearly forgotten because of its limited range of applicability until Fichera [6] independently rediscovered it and made decisive improvements, which will be discussed in Section 6.

Let K be a negative-definite integral operator defined by a given kernel $k(s, t)$, where s and t are variables in one or more dimensions. We wish to determine lower bounds for the eigenvalues $\mu_1 \leq \mu_2 \leq \cdots < 0$ of the equation

$$\int k(s, t)u(t)\,dt = \mu u(s)$$

or, more simply,

$$Ku = \mu u.$$

Here, μ denotes the reciprocal of Fredholm's λ. A finite number of upper bounds $\mathrm{M}_1 \leq \mathrm{M}_2 \leq \cdots \leq \mathrm{M}_r$ may be found by the Rayleigh–Ritz Method.

To obtain lower bounds, we use one of the traces

$$J_{2q}(k) = \int k^{(2q)}(s, s)\,ds = \sum_{i=1} \mu_i^{2q} \qquad (q = 1, 2, \ldots),$$

where $k^{(2q)}(s, t)$ is the iterated kernal of index $2q$. Therefore, we have

$$J_{2q}(k) \geq \mu_n^{2q} + \sum_{i \neq n} \mu_i^{2q},$$

where the sum is taken over a *finite number* of *any* indices i, $i \neq n$. Recalling that $\mu_i \leq \mathrm{M}_i < 0$, we have

$$J_{2q}(k) \geq \mu_n^{2q} + \sum_{i \neq n} \mathrm{M}_i^{2q}. \tag{1}$$

† This paper is sometimes confused with another paper by Trefftz which deals with lower bounds for the Dirichlet integral and not with eigenvalues [12a, p. 146].

In this way, a computable lower bound for μ_n is given by

$$-\left\{J_{2q}(k) - \sum_{i \neq n} M_i^{2q}\right\}^{1/2q} \leq \mu_n. \tag{2}$$

Let us note in passing that Trefftz only considered the case $q = 1$.

It is clear from the above discussion that this method can be applied only to negative (or positive) operators for which the trace or a higher trace exists and is known. This assumption is not fulfilled in quantum theory where methods of intermediate problems are applicable.

6. Fichera's Construction of Intermediate Green's Operators

Trefftz himself pointed out that his procedure has a very limited range of applicability in the area of partial differential equations, since the kernel of the integral equation, namely the Green's function $g(x, y)$, is rarely known in a usable form, except maybe for a circular domain. Even in this case, its determination is far from being trivial, for instance, in the problem of the vibration of a circular plate clamped on the edge, (see [DFFS1]), where the computations are performed by means of this method.

Decisive progress was achieved by Fichera for elliptic operators of any order on a bounded domain Ω, at least in the case in which the solution and an appropriate number of its derivatives *vanish on the boundary*. Fichera showed that, instead of the Green's function, it is sufficient to know explicitly the fundamental solution $s(x, y)$ for the differential operator and that the fundamental solution leads to an explicit representation of *intermediate Green's functions* $g_1(x, y), g_2(x, y), \ldots$ satisfying

$$J_{2q}(g_1) \geq J_{2q}(g_2) \geq \cdots \geq J_{2q}(g). \tag{1}$$

Therefore, the functions g_i can be used in place of g in inequality (5.2) and yield lower bounds.

To achieve this end, a new representation of Green's operator is derived. We limit ourselves here to an important case for applications, namely the biharmonic operator in a plane domain Ω.

Let us again consider the equation of the vibrating plate

$$\Delta\,\Delta w - \mu w = 0,$$

with w and its normal derivative $\partial w/\partial n$ vanishing on the boundary $\partial\Omega$. We replace these boundary conditions, as in Section 4.14, by the conditions $w = 0$ on the boundary and Δw orthogonal to every harmonic function.

Then, by Eq. (4.14.4), we have to solve

$$-GGu + PGGu = \lambda u, \qquad Pu = 0, \tag{2}$$

where P is the orthogonal projection operator onto the subspace of harmonic functions.

In order to find the Green's function for the biharmonic equation, we reintroduce w in (2) by putting $w = Gu$, to obtain

$$-Gw + PGw = \lambda u. \tag{3}$$

Applying the operator G on both sides of (3), we finally have

$$-GGw + GPGw = \lambda w. \tag{4}$$

This eigenvalue equation for w leads to the conjecture that the Green's operator Γ for the biharmonic problem should be given by

$$\Gamma = -GG + GPG. \tag{5}$$

In order to prove (5), we first observe that $G = S + T$, where S is an operator defined by the fundamental solution [with a logarithmic kernel $s(x, y)$], namely

$$[Sf](x) = \int_\Omega s(x, y) f(y)\, dy \tag{6}$$

and T is an integral operator with a harmonic kernel. Noting that $(I - P)T = 0$ and $TP = T$, we can rewrite Γ as

$$\begin{aligned} \Gamma &= -(S + T)(I - P)(S + T) = -(S + T)(I - P)S \\ &= -S(I - P)S - TS + TS = -SS + SPS. \end{aligned} \tag{7}$$

Using this form, we show that Γ is the Green's operator for the biharmonic operator, namely that, for an arbitrary $f \in \mathscr{L}^2(\Omega)$, the problem

$$\Delta\, \Delta v = -f \qquad \text{in} \quad \Omega,$$

$$v = \partial v / \partial n = 0 \qquad \text{on} \quad \partial\Omega$$

admits the solution

$$v = \Gamma f.$$

To this end, we note that

$$P[Sf - PSf] = 0. \tag{8}$$

This means that $\phi(y) = -Sf + PSf$ is orthogonal to any harmonic function in Ω. Let us consider a circular disk Ω' containing $\overline{\Omega}$ (the closure of Ω) and let $\Omega' - \overline{\Omega}$ be the "belt" bounded by the boundaries of Ω and Ω'. (We assume that Ω is simply connected.) We denote by $\psi(x)$ any function in $\mathscr{L}^2(\Omega' - \Omega)$. Then, the function

$$\int_{\Omega'-\Omega} s(x, y)\psi(x)\, dx$$

is harmonic in Ω and therefore, by (8), we have

$$\int_{\Omega} \phi(y) \int_{\Omega'-\Omega} s(x, y)\psi(x)\, dx\, dy = 0.$$

Interchanging the order of integration (Fubini's theorem [26]), we see that the function

$$\int_{\Omega} s(x, y)\phi(y)\, dy = S\phi \tag{9}$$

is orthogonal to any $\mathscr{L}^2$ function in $\Omega' - \Omega$ and, therefore, it is almost everywhere zero in $\Omega' - \Omega$. As this integral (9) is a logarithmic potential and, therefore, continuously differentiable in Ω', it is identically zero in $\Omega' - \Omega$, so that $S\phi$ and its normal derivative vanish on the boundary of Ω. This means that

$$S\phi = -SSf + SPSf$$

satisfies the boundary conditions of the biharmonic Green's function. It is easy to see that $S\phi$ satisfies

$$\Delta\, \Delta S\phi = -f.$$

In fact,

$$\Delta\, \Delta(-SSf + SPSf) = \Delta(-Sf + PSf) = -\Delta Sf = -f.$$

Since f is arbitrary, we have found that Γ is the Green's operator for the biharmonic equation.

While formula (5) is a representation of the Green's operator, it does not lead to an explicit formula for the Green's function $g(x, y)$ since the projector P is not explicitly known for an arbitrary domain Ω. Nevertheless, (8) leads to the explicit construction of *intermediate Green's functions* in the following way.

Let $p_1, p_2, \ldots$ be a linearly independent sequence of $\mathscr{L}^2$ functions such that $\Delta p_i = 0$ in Ω $(i = 1, 2, \ldots)$; let $\mathfrak{P}_n = \mathrm{sp}\{p_1, p_2, \ldots, p_n\}$; and let P_n be the orthogonal projection operator onto $\mathfrak{P}_n$. We put $\Gamma_n = -SS + SP_n S$ ($n = 1, 2, \ldots$). Then, for any index n, we have

$$\begin{aligned}(\Gamma u, u) &= ([-SS + SPS]u, u) \\ &= (-SSu, u) + (PSu, Su) \\ &\geq (-SSu, u) + (P_n Su, Su) \\ &= ([-SS + SP_n S]u, u) = (\Gamma_n u, u).\end{aligned} \tag{10}$$

This inequality is fundamental for the computation of lower bounds, since Γ_n is an explicitly known integral operator. In fact, in view of the definition of S in Eq. (9), operator SS can be expressed as an integral operator, its kernel being a convolution. Moreover, the degenerate operator $SP_n S$ can also be expressed as an integral operator.

Therefore, we have explicitly computable (through quadratures) *intermediate Green's functions* $g_n(s, t)$ corresponding to the operators Γ_n ($n = 1, 2, \ldots$). Moreover, since we have inequalities (1), such functions $g_n(x, y)$ may be used in place of $g(x, y)$ in inequality (5.1), thereby yielding a sequence of nondecreasing lower bounds of the type given in inequality (5.2).

It is clear that in Fichera's procedure $-SS$ corresponds to a base operator and Γ_n corresponds to an intermediate operator. However, unlike intermediate problems, only the *traces* and not the individual eigenelements of the base and intermediate operators appear.

This method has been applied numerically to a number of problems with *vanishing boundary conditions*. For the vibrating square plate excellent lower bounds were computed for the first 45 eigenvalues. For instance, Fichera [F7] obtained the inequality

$$13.29376 \leq \lambda_1 \leq 13.29378.$$

For the Mathieu equation (5.5.2) with $s = 8$, Fichera gave lower bounds for 10 eigenvalues. Fichera and his collaborators [DFFS1] computed lower bounds for the first 160 eigenvalues of the vibrating, clamped circular plate. In this case, the Green's function was explicitly known, so that it was not necessary to construct intermediate Green's functions.

Chapter Seven

The New Maximum-Minimum Theory

1. The Basis of Weinstein's New Maximum–Minimum Theory

It is remarkable that Weyl's first lemma and intermediate problems of the first type have nearly the same formulation. In fact, in both cases we have to consider the minimum $\lambda(p_1, p_2, \ldots, p_{n-1})$ of $R(u) = (Au, u)/(u, u)$ for $u \in \mathfrak{D}$ subject to orthogonality conditions

$$(u, p_i) = 0 \qquad (i = 1, 2, \ldots, n-1). \tag{1}$$

One essential difference, however, is that in Weyl's problem the vectors p_i are completely arbitrary elements in $\mathfrak{H}$ (not necessarily independent), whereas in intermediate problems the p_i have to be selected from a given subspace $\mathfrak{P} \subset \mathfrak{H}$.

This obvious connection remained unnoticed for many years, even in Weyl's famous Gibb's lecture [W32] in which his maximum–minimum theory and the first type of intermediate problems are discussed side by side. This connection became the basis of *Weinstein's new maximum–minimum theory*. See for instance [W16, W18, W22, S11].

As was already indicated in Section 3.2, this analogy shows that $\lambda(p_1, p_2, \ldots, p_{n-1})$ is the first eigenvalue of

$$Au - P_r Au = \lambda u, \qquad P_r u = 0 \tag{2}$$

or the first nontrivial eigenvalue of

$$Q_r A Q_r u = \lambda u, \tag{3}$$

for some r $(0 \le r \le n-1)$, r being the number of independent vectors p_i.

Therefore, the methods of Chapter 4 can be applied to the determination of $\lambda(p_1, p_2, \ldots, p_{n-1})$. Of course, in the case of the maximum–minimum theory, the eigenvalue problems (2–3) are, strictly speaking, not intermediate problems since we do not have a "given" problem, say $QAQu = \lambda u$. However, in order to simplify the terminology, we call (2) or (3) the "$(n-1)$th intermediate problem" to indicate dependence on $n-1$ constraints and we use the same notations as in intermediate problems, namely $\lambda_j^{(n-1)}$, $u_j^{(n-1)}$, etc.

The new maximum–minimum theory which we now develop gives necessary and sufficient conditions under which the equality

$$\lambda(p_1, p_2, \ldots, p_{n-1}) = \lambda_n^{(0)}$$

holds. Moreover, when A is compact the new theory provides a completely new and independent proof of Weyl's inequality (3.1.2).

In the following, A will be assumed to be in class $\mathscr{S}$, which of course includes compact operators.

Let us first discuss a nearly trivial case [W16] which does not require the more elaborate techniques of the following sections.

Lemma 1. *If $\lambda_1 = \lambda_2 = \cdots = \lambda_n$, then $\lambda(p_1, p_2, \ldots, p_{n-1}) = \lambda_n$ for any choice of $p_1, p_2, \ldots, p_{n-1}$.*

REMARK. A similar statement applies to higher eigenvalues of multiplicity greater than one.

PROOF. This follows immediately from the old maximum–minimum theory, since $\lambda_1 = \lambda_n$. Let us also give here a second proof which has a curious sidelight. In fact, the multiplicity of λ_1 $(=\lambda_n)$ is obviously $\geq n$. Let

$$u_0 = \alpha_1 u_1 + \alpha_2 u_2 + \cdots + \alpha_n u_n \neq 0$$

satisfy $n-1$ conditions (1). Then, u_0 is an eigenfunction of (3) corresponding to the eigenvalue λ_1 $(=\lambda_n)$, since we have

$$Q_r A Q_r u_0 = Q_r A u_0 = Q_r(\lambda_1 u_0) = \lambda_1 u_0 .$$

As $\lambda_1 = \lambda_n$, λ_n is the lowest eigenvalue of (3), which means that

$$\lambda(p_1, p_2, \ldots, p_{n-1}) = \lambda_n$$

for any choice of $p_1, p_2, \ldots, p_{n-1}$.

Notice that we used Weyl's test function u_0 as a *solution* of (3) and not for estimating an upper bound to $R(u)$.

2. The Case of One Constraint

We start by considering the problem in which we have only one constraint given by an arbitrary element in $\mathfrak{H}$, $p = p_1$, with $(p, p) = 1$. We have to determine the minimum of $R(u)$ for u orthogonal to p, which is the lowest eigenvalue $\lambda_1^{(1)}$ of

$$Au - (Au, p)p = \lambda u, \qquad (p, u) = 0. \tag{1}$$

We introduce the following notations. Let $N_k = N_k(\lambda_n^{(0)})$ $(k = 0, 1)$, denote the number of eigenvalues $\lambda^{(k)}$ that are less than or equal to $\lambda_n^{(0)}$. We denote by $\mu_0(\lambda_n^{(0)})$ the multiplicity of $\lambda_n^{(0)}$ in the base problem and by $\mu_1(\lambda_n^{(0)})$ the multiplicity of $\lambda_n^{(0)}$ in the first intermediate problem (1). We again consider the function

$$W_{01}(\lambda) \equiv (R_\lambda p, p). \tag{2}$$

We recall from Section 4.8 that $W_{01}(\lambda)$ is positive for large, negative λ and increases with λ. This is an important fact which will be used throughout the following discussion.

We now prove the following result.

Theorem 1. *For an arbitrary choice of p, we have*

$$\lambda_{n-1}^{(1)} \leq \lambda_n^{(0)} \qquad (n = 2, 3, \ldots). \tag{3}$$

Moreover, we have

$$\lambda_{n-1}^{(1)} = \lambda_n^{(0)} \tag{4}$$

if and only if, for all admissible $\varepsilon > 0$,

$$W_{01}(\lambda_n^{(0)} - \varepsilon) > 0, \tag{5}$$

and at the same time

$$\lambda_{n-1}^{(0)} = \lambda_n^{(0)} \tag{6}$$

or

$$W_{01}(\lambda_n^{(0)} - \varepsilon) < 0, \tag{7}$$

in which case (6) *is not required to hold.*

REMARK 1. Here, "admissible ε" means the following. The function $W_{01}(\lambda)$ has at most a finite number of zeros and a finite number of poles that are less than λ_n. We put λ_0 equal to the largest such zero or pole (or equal to any number less than λ_n if there are no zeros or poles). Then, if the positive number ε is such that $\lambda_0 < \lambda_n - \varepsilon$, we say that ε is admissible. Obviously, any sufficiently small $\varepsilon > 0$ is admissible.

Proof of Theorem. In the case $\lambda_1^{(0)} = \lambda_2^{(0)} = \cdots = \lambda_n^{(0)}$ $(n \geq 2)$, we see from Lemma 1.1 that equality (4) holds for any choice of p (even $p = 0$). Therefore, we will assume once and for all that, for a certain $m \geq 2$,

$$\lambda_{m-1}^{(0)} < \lambda_m^{(0)} = \lambda_{m+1}^{(0)} = \cdots = \lambda_n^{(0)} \qquad (2 \leq m \leq n). \tag{8}$$

Therefore, $N_0 - \mu_0 + 1 = m$. Statement (3) contains the first part of the fundamental lemma for $n = 2$, namely

$$\lambda_1^{(1)} = \lambda(p) \leq \lambda_2. \tag{9}$$

We first show that conditions (5) and (6) or condition (7) is sufficient. For the proof, we will consider several cases. These cases will be illustrated in a series of diagrams, in which, for simplicity, we put $n = 2$. Let us note again that $W_{01}(\lambda_n^{(0)})$, or, more concisely, $W(\lambda_n^{(0)})$, is either infinite, positive, zero, or negative.

Case A. Assume $W(\lambda_n^{(0)}) = \infty$, which implies $W(\lambda_n^{(0)} - \varepsilon) > 0$. We show that $N_1(\lambda_n^{(0)}) = N_0(\lambda_n^{(0)}) - 1$, or, more concisely, $N_1 = N - 1$. In fact, by (4.8.1), one eigenvalue $\lambda_n^{(0)}$ is lost from the spectrum of the first intermediate problem. By this, we mean

$$\mu_1(\lambda_n^{(0)}) = \mu_0(\lambda_n^{(0)}) - 1. \tag{10}$$

If in going to the left, we find an index $v < n$ for which $W(\lambda_v^{(0)}) = \infty$, then, by the general theory, there will be a new eigenvalue $\lambda^{(1)}$ $(\lambda_v^{(0)} < \lambda^{(1)} < \lambda_n^{(0)})$ the position of which is determined by $W(\lambda^{(1)}) = 0$ (see Fig. 2, in which, as in all figures, $\lambda_1^{(0)}$ plays the role of $\lambda_v^{(0)}$). This eigenvalue compensates the loss of one $\lambda_n^{(0)}$. This reasoning can be repeated. However, if, in going to the left, we come to the smallest $\lambda_\alpha^{(0)}$ for which $W(\lambda_\alpha^{(0)}) = \infty$ $(1 \leq \alpha \leq v)$, then the loss of $\lambda = \lambda_\alpha^{(0)}$ is not compensated. It may also happen, however, that $W(\lambda)$ remains finite for $\lambda < \lambda_n^{(0)}$ (see Fig. 1). Then index n plays the role of α, so that again the loss of $\lambda_n^{(0)}$ is not compensated. Thus, we see that if $W(\lambda_n^{(0)}) = \infty$, we always have

$$N_1(\lambda_n^{(0)}) = N_0(\lambda_n^{(0)}) - 1. \tag{11}$$

Therefore, in the first intermediate problem, there are $N_1 = N - 1$ eigenvalues that are less than or equal to $\lambda_n^{(0)}$:

$$\lambda_{m-1}^{(1)} < \lambda_m^{(1)} = \cdots = \lambda_{n-1}^{(1)} = \cdots = \lambda_{N-1}^{(1)} \leq \lambda_n^{(0)}, \tag{12}$$

while in the base problem, we have

$$\lambda_{m-1}^{(0)} < \lambda_m^{(0)} = \cdots = \lambda_n^{(0)} = \cdots = \lambda_N^{(0)}. \tag{13}$$

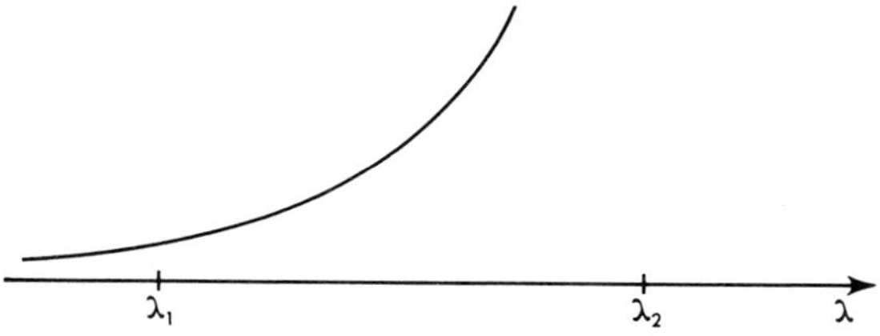

FIG. 1

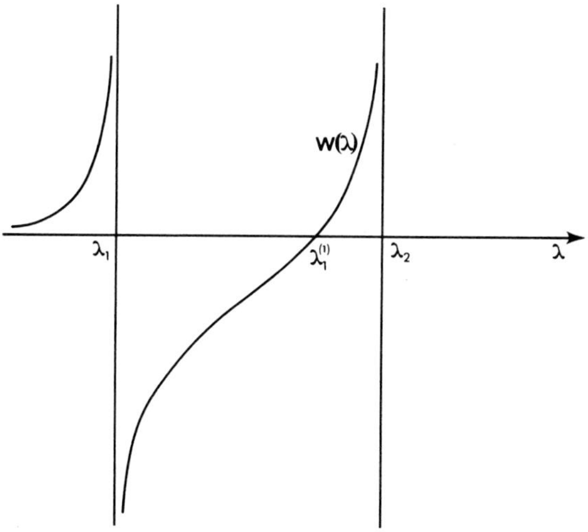

FIG. 2

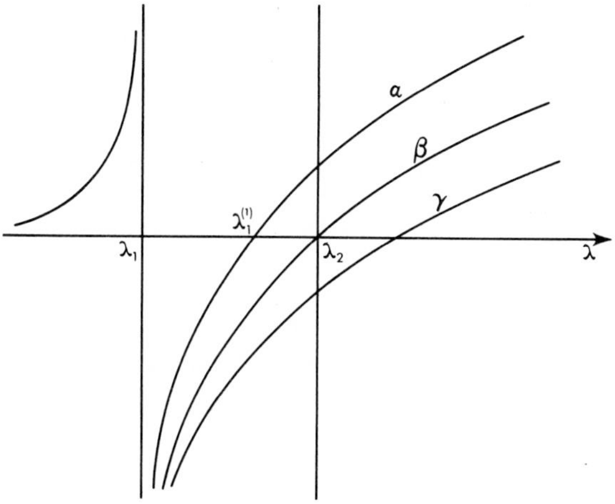

FIG. 3

From (12), we see that (3) holds, and, upon comparison of (12) and (13), we conclude that (4) holds if and only if $n > m$. In other words, the necessary and sufficient condition for the equality $\lambda_{n-1}^{(1)} = \lambda_n^{(0)}$ in the present case is (6).

Case B. Assume $W(\lambda_n^{(0)})$ is positive but finite (see Fig. 3, curve α), which implies that $W(\lambda_n^{(0)} - \varepsilon)$ is also positive. In this case, $N_1(\lambda_n^{(0)}) = N_0(\lambda_n^{(0)})$ or, more concisely, $N_1 = N$. In fact, by (4.8.1), $\lambda_n^{(0)}$ is an eigenvalue of the same multiplicity. If $W(\lambda_n^{(0)})$ is positive and $W(\lambda)$ remains positive for all $\lambda \leq \lambda_n^{(0)}$ (see Fig. 1), then we have

$$\lambda_i^{(1)} = \lambda_i^{(0)} \qquad (i = 1, 2, \ldots, N) \tag{14}$$

and therefore $N_1 = N$. If, however, we have $W(\lambda_n^{(0)})$ positive and $W(\lambda_\nu^{(0)}) = \infty$ for $\nu < n$ (see Fig. 3), then in the spectrum of the first intermediate problem we have the eigenvalue $\lambda_n^{(0)}$ with the multiplicity μ_0 and another eigenvalue $\lambda^{(1)}$ satisfying

$$\lambda_\nu^{(0)} < \lambda^{(1)} < \lambda_n^{(0)}. \tag{15}$$

Of course, this $\lambda^{(1)}$ could be numerically equal to one of the $\lambda^{(0)}$'s that lie between $\lambda_\nu^{(0)}$ and $\lambda_n^{(0)}$. By case A, we know that in the interval $\lambda \leq \lambda_\nu^{(0)}$ we lose an eigenvalue when we go to the first intermediate problem, which compensates the gain given by (15). Therefore, we again have $N_1 = N$. Thus, in case B, we have exactly N eigenvalues satisfying

$$\lambda_1^{(1)} \leq \lambda_2^{(1)} \leq \cdots \leq \lambda_{n-1}^{(1)} \leq \cdots \leq \lambda_N^{(1)} \quad (\leq \lambda_n^{(0)}), \tag{16}$$

which yields (3). As the multiplicity of $\lambda_n^{(0)}$ is preserved, we have

$$\lambda_{m-1}^{(1)} < \lambda_m^{(1)} = \lambda_{m+1}^{(1)} = \cdots = \lambda_n^{(1)} = \cdots = \lambda_N^{(1)} \quad (= \lambda_n^{(0)}), \tag{17}$$

where $m = N - \mu_0 + 1$. These eigenvalues are equal to each of the μ_1 $(= \mu_0)$ eigenvalues

$$\lambda_m^{(0)} = \cdots = \lambda_n^{(0)} = \cdots = \lambda_N^{(0)}. \tag{18}$$

Therefore, comparing (17) with (18), we have, as in case A, the equality $\lambda_{n-1}^{(1)} = \lambda_n^{(0)}$ if and only if $\lambda_{n-1}^{(0)} = \lambda_n^{(0)}$.

Case C. Assume $W(\lambda_n^{(0)}) = 0$ (see Fig. 3, curve β), which implies that $W(\lambda_n^{(0)} - \varepsilon) < 0$. Then, there is a $\lambda_\nu^{(0)} < \lambda_n^{(0)}$ such that $W(\lambda_\nu^{(0)}) = \infty$. By (4.8.1), we obtain $\mu_1(\lambda_n^{(0)}) = \mu_0(\lambda_n^{(0)}) + 1$ and the gain in the number of eigenvalues is compensated by a single loss in the number of eigenvalues $\lambda^{(1)}$ that occur in the interval $\lambda \leq \lambda_\nu^{(0)}$. Therefore, N_1 is equal to N and hence we have

$$\lambda_{m-1}^{(0)} < \lambda_m^{(0)} = \cdots = \lambda_n^{(0)} = \cdots = \lambda_N^{(0)} \qquad (2 \leq m \leq n) \tag{19}$$

and

$$\lambda_{m-1}^{(1)} = \lambda_m^{(1)} = \cdots = \lambda_{n-1}^{(1)} = \cdots = \lambda_N^{(1)} \quad (=\lambda_n^{(0)}), \tag{20}$$

which yields (3) and the equality $\lambda_{n-1}^{(1)} = \lambda_n^{(0)}$ for all n $(m \le n \le N)$.

CASE D. Assume $W(\lambda_n^{(0)}) < 0$ (see Fig. 3, curve γ), which implies $W(\lambda_n^{(0)} - \varepsilon) < 0$. Then we have, by (4.8.1), $\mu_1 = \mu_0$. Furthermore, there is a $\lambda_\nu^{(0)} < \lambda_n^{(0)}$ for which $W(\lambda_\nu^{(0)}) = \infty$ and $W(\lambda) < 0$ for $\lambda_\nu^{(0)} < \lambda < \lambda_n^{(0)}$. Therefore, there is no loss of eigenvalues in the interval $\lambda_\nu^{(0)} < \lambda \le \lambda^{(0)}$, but an eigenvalue is lost in the interval $\lambda \le \lambda_\nu^{(0)}$. Hence, $N_1 = N - 1$. Therefore, we have

$$\lambda_{m-1}^{(0)} < \lambda_m^{(0)} = \cdots = \lambda_n^{(0)} = \cdots = \lambda_N^{(0)} \tag{21}$$

and

$$\lambda_{m-1}^{(1)} = \cdots = \lambda_{n-1}^{(1)} = \cdots = \lambda_{N-1}^{(1)} \quad (=\lambda_n^{(0)}), \tag{22}$$

which yields inequality (3) and equality (4), $\lambda_{n-1}^{(1)} = \lambda_n^{(0)}$ $(m \leqq n \leqq N)$.

Thus, the sufficiency of conditions (5) and (6) or condition (7) is proved. To prove necessity, we assume $W(\lambda_n^{(0)} - \varepsilon) > 0$ and $m = n$. Then, by the discussion of cases A and B, we see that the equality sign in (3) cannot hold. Therefore, for (4) to hold, we have necessarily

$$W(\lambda_n^{(0)} - \varepsilon) < 0 \tag{23}$$

or

$$W(\lambda_n^{(0)} - \varepsilon) > 0$$

and $m < n$ in the second case, which completes the proof of the theorem.

REMARK 2. Let us note here that in the case $m = n$, condition (7) implies that $W(\lambda_n^{(0)})$ must be finite and we could replace condition (7) by the equivalent condition $W(\lambda_n^{(0)}) \le 0$.

The following table covers all cases. In the last line, we list all indices

WEINSTEIN'S TABLE

FIP	$W \equiv 0$	$m, m+1, \ldots, n, \ldots, N-1, N$
FIP	$W = \infty$	$m, m+1, \ldots, n, \ldots, N-1$
FIP	$W > 0$	$m, m+1, \ldots, n, \ldots, N-1, N$
FIP	$W = 0$	$m-1, m, m+1, \ldots, n, \ldots, N-1, N$
FIP	$W < 0$	$m-1, m, m+1, \ldots, n, \ldots, N-1$
BP		$m, m+1, \ldots, n, \ldots, N-1, N$

k for which $\lambda_k^{(0)}$ [that is, eigenvalue λ_k of the base problem (BP)] is numerically equal to $\lambda_n^{(0)}$. The other lines give the indices of the eigenvalues $\lambda_k^{(1)}$ [that is, eigenvalues of the first intermediate problem (FIP)] which are numerically equal to $\lambda_n^{(0)}$. On the top line is the case (excluded up to now) $p = 0$, $W \equiv 0$. The remaining lines correspond to the four possible values of $W(\lambda)$ for $\lambda = \lambda_n^{(0)} - \varepsilon$ and $W \not\equiv 0$.

From this table, we see that $\lambda_{m-1}^{(1)} = \lambda_n^{(0)}$ if and only if $W(\lambda_n^{(0)} - \varepsilon)$ is negative. Furthermore, the line corresponding to the case $W \equiv 0$ is identical to that for $W > 0$ as well as to the line corresponding to the base problem.

3. The Case of Orthonormal Constraints

The object of this section is to give a new proof of the classical statement

$$\lambda(p_1, p_2, \ldots, p_{n-1}) \leq \lambda_n \tag{1}$$

and to establish a necessary and sufficient condition for the equality sign to hold. In the trivial case $p_1 = p_2 = \cdots = p_{n-1} = 0$, inequality (1) is obviously true since it simply states that $\lambda_1 \leq \lambda_n$. For the moment, we assume that $(p_i, p_k) = \delta_{ik}$ $(i, k = 1, 2, \ldots, r, 1 \leq r \leq n-1)$, and $p_{r+1} = p_{r+2} = \cdots = p_{n-1} = 0$. We denote the principal minors of $W_{0r}(\lambda)$ by $W_1, W_2, \ldots, W_r$, where $W_1 = W_{01}$ is the function just considered in Section 2. Since each of these functions is meromorphic for $\lambda < \lambda_\infty$, we can, in a way similar to that in Section 2, put λ_0 equal to the largest zero or pole of $W_1, W_2, \ldots, W_r$ that is less than λ_n and again define an admissible ε to be any $\varepsilon > 0$ such that $\lambda_0 < \lambda_n - \varepsilon$.

We can now state the following result.

Theorem 1. *For p_i satisfying the above conditions, we have*

$$\lambda(p_1, p_2, \ldots, p_{n-1}) \leq \lambda_n. \tag{2}$$

A necessary condition for equality to hold in (2) *is that $r \geq m - 1$. If this condition is satisfied, then a necessary and sufficient condition for equality to hold is the following: For any admissible ε, either* (a) *if $W_1(\lambda_n - \varepsilon)$ is negative, the sequence*

$$W_1, \quad W_2, \ldots, W_{r-1}, \quad W_r \tag{3}$$

evaluated at $\lambda = \lambda_n - \varepsilon$ alternates in sign exactly $(m-2)$ times [in other words, the sequence changes sign with exactly $(r - m + 1)$ exceptions which may occur at any place]; *or* (b) *if $W_1(\lambda_n - \varepsilon)$ is positive, sequence* (3) *alternates in sign exactly $(m-1)$ times* [in other words, the sequence changes sign with exactly $(r - m)$ exceptions which may occur at any place].

If $m = n = 2$, there is only one element in the sequence (3) and this element $W_1 = W_{01}$ is negative. This may be considered as a special case of (a). The proof of the theorem is based upon repeated application of the results of Section 2.

PROOF. If we consider the first intermediate problem as the base problem for the second intermediate problem, then the function $W_{12}(\lambda)$ (see Section 4.8) plays the role of W_{01} of Section 2. In a similar way, we have $W_{i,\,i+1}$, which is used in the transition from the ith intermediate problem to the $(i+1)$th intermediate problem $(i = 0, 1, 2, \ldots, r-1)$. Of course, $W_{i,\,i+1}$ is identically zero for $i = r, r+1, \ldots, n-2$.

A repeated application of Theorem 2.1 leads to

$$\lambda_1^{(n-1)} \leq \lambda_2^{(n-2)} \leq \cdots \leq \lambda_{n-r}^{(r)} \leq \cdots \leq \lambda_{n-1}^{(1)} \leq \lambda_n^{(0)} = \lambda_n, \tag{4}$$

which proves (2). Clearly, the equality $\lambda_1^{(n-1)} = \lambda_n^{(0)}$ holds if and only if equality holds throughout (4).

Let us show now that if equality holds in (4), we must then have $r \geq m - 1$. To this end, let us denote by m_i the smallest index such that in the ith intermediate problem $\lambda_{m_i}^{(i)} = \lambda_n^{(0)}$. As we have equality throughout (4), it follows that $1 \leq m_i \leq n - i$. In particular, $m_{n-1} = 1$. On the other hand, by the diagram at the end of Section 2 (obviously valid for two successive intermediate problems), we have either

$$m_i = m_{i-1} \tag{5}$$

or

$$m_i = m_{i-1} - 1, \tag{6}$$

where $m_0 = m$. This shows that $m - r \leq m_r \leq m$. Moreover, in view of the case $W \equiv 0$, we see that $m_r = m_{r+1} = \cdots = m_{n-1}$. Hence, $m - r \leq m_r = m_{n-1} = 1$, which shows that $r \geq m - 1$.

Assuming $r \geq m - 1$, we will now show that a necessary and sufficient condition for the equality $\lambda_1^{(n-1)} = \lambda_n^{(0)}$ to hold is that (6) occurs exactly $(m - 1)$ times, such occurrences being possible at any place in the sequence of problems.

In fact, suppose $\lambda_1^{(n-1)} = \lambda_n^{(0)}$. Then, $m_{n-1} = m - s_{n-1}$, where s_i denotes the number of times that (6) holds in the transition from the base problem to the ith intermediate problem. But, since $m_{n-1} = 1$, it follows that $s_{n-1} = m - 1$, which establishes the necessity of the condition. To prove sufficiency, suppose, to the contrary, that $\lambda_1^{(n-1)} < \lambda_n^{(0)}$. Then, there exists an index k such that

$$\lambda_{n-k-1}^{(k+1)} < \lambda_{n-k}^{(k)} = \lambda_{n-k+1}^{(k-1)} = \cdots = \lambda_{n-1}^{(1)} = \lambda_n^{(0)}, \tag{7}$$

where $0 \leq k < n-1$. In particular, by the table of Section 2, the inequality

$$\lambda_{n-k-1}^{(k+1)} < \lambda_{n-k}^{(k)} \tag{8}$$

implies that $n - k = m_k$ and $W_{k,k+1}(\lambda)$ is either identically zero or positive at $\lambda = \lambda_n - \varepsilon$. In every other case, we have equality in (8). Furthermore, we have $m_k = m - s_k$ as well as $m_k = n - k$. Hence, we see that $s_k = m - n + k$. Therefore, in the transition from the base problem to the kth intermediate problem, Eq. (6) holds $k - s_k = n - m$ times. By assumption, Eq. (6) holds $(m - 1)$ times in the transition from the base problem to the $(n-1)$th intermediate problem. Therefore, in the remaining $(n - 1 - k)$ steps from the kth problem to $(n-1)$th intermediate problem, Eq. (6) holds $(m - 1 - s_k) = (n - 1 - k)$ times. This implies $W_{k,k+1}(\lambda_{n-k}^{(k)} - \varepsilon) = W_{k,k+1}(\lambda_n - \varepsilon)$ is negative, a contradiction establishing the sufficiency of our condition. It is now obvious that we can recapitulate the above results in the following statement.

Criterion. *A necessary and sufficient condition for the equality* $\lambda_1^{(n-1)} = \lambda_n^{(0)}$ $(=\lambda_n)$ *to hold is that, in the sequence*

$$W_{01}, W_{12}, \ldots, W_{r-1,r} \qquad (r \geq m-1) \tag{9}$$

evaluated at $\lambda_n - \varepsilon$, *there are exactly* $(m-1)$ *elements for which* $W_{h-1,h}$ *is negative. The remaining* $(r - m + 1)$ *elements are positive.*

To complete the proof of our theorem, which expresses the criterion in terms of the base problem alone, we again use the decomposition (see Section 4.8)

$$W_h = W_{01} W_{12} \cdots W_{h-1,h} \qquad (h = 1, 2, \ldots, r). \tag{10}$$

Hence, if $W_{01}(\lambda_n - \varepsilon)$ is negative and if there are $(m-1)$ elements in (9) that are negative, then the sequence (3) alternates in sign exactly $(m-2)$ times. Similarly, if $W_{01}(\lambda_n - \varepsilon)$ is positive, then the sequence (3) alternates in sign $(m-1)$ times, which completes the proof of the theorem.

4. The General Formulation

As the minimum $\lambda(p_1, p_2, \ldots, p_{n-1})$ depends only on the subspace $\mathfrak{P}_{n-1} = \mathrm{sp}\{p_1, p_2, \ldots, p_{n-1}\}$, we can now formulate our criterion in such a way that it will be independent of the basis chosen.

We again let $r = \dim \mathfrak{P}_{n-1}$ and from now on let $p_1, p_2, \ldots, p_r$ be any basis (not necessarily orthonormal) of $\mathfrak{P}_{n-1}$ and let $p_1', p_2', \ldots, p_r'$ be an orthonormal basis for $\mathfrak{P}_{n-1}$. There exists a matrix $\{\alpha_{ik}\}$ connecting these bases by the formulas

$$p_i = \sum_{s=1}^{r} \alpha_{is} p_s' \qquad (i = 1, 2, \ldots, r).$$

Let us denote by $a_{ik}(\lambda)$ and $a_{ik}'(\lambda)$ the elements of $W_r(\lambda)$ formed with respect to the two bases and consider the corresponding quadratic forms

$$\sum_{i,k=1}^{r} a_{ik}(\lambda)\xi_i \xi_k$$

and

$$\sum_{i,k=1}^{r} a_{ik}'(\lambda)\xi_i' \xi_k'.$$

If we put

$$\xi_s' = \sum_{i=1}^{r} \alpha_{is} \xi_i \qquad (s = 1, 2, \ldots, r),$$

we see that the elements $a_{ik}(\lambda)$ are transformed into $a_{ik}'(\lambda)$ by the same rule as the coefficients of a quadratic form. With this in mind, we can now formulate the main theorem.

Theorem 1. (*Weinstein's new maximum–minimum principle*) *For any choice of* $p_1, p_2, \ldots, p_{n-1} \in \mathfrak{H}$, *we have the inequality*

$$\lambda(p_1, p_2, \ldots, p_{k-1}) \leq \lambda_n. \tag{1}$$

A necessary condition for the equality

$$\lambda(p_1, p_2, \ldots, p_{n-1}) = \lambda_n \tag{2}$$

to hold is that the dimension r of the subspace $\mathfrak{P}_{n-1}$ *generated by* $p_1, p_2, \ldots, p_{n-1}$ *satisfy the inequality*

$$m - 1 \leq r, \tag{3}$$

where $m = \min\{j \mid \lambda_j = \lambda_n\}$. *Let* $p_1, p_2, \ldots, p_r$ *denote a basis for* $\mathfrak{P}_{n-1}$ (*reindexing if necessary to eliminate superfluous vectors*). *Assuming that* (3) *holds, we have the following necessary and sufficient condition for* (2): *For any admissable* $\varepsilon > 0$, *the quadratic form defined by the symmetric matrix*

$$\{(R_\lambda p_i, p_k)\} \qquad (i, k = 1, 2, \ldots, r)$$

evaluated at $\lambda_n - \varepsilon$ *has in canonical coordinates* $\mathbf{x}_1, \mathbf{x}_2, \ldots, \mathbf{x}_r$ *the form*

$$-\mathbf{x}_1^2 - \mathbf{x}_2^2 - \cdots - \mathbf{x}_{m-1}^2 + \mathbf{x}_m^2 + \mathbf{x}_{m+1}^2 + \cdots + \mathbf{x}_r^2. \tag{4}$$

PROOF. The new proof of inequality (1) as well as the proof of necessary condition (3) have already been given in the previous section. We now show that the necessary and sufficient condition given here is equivalent to the one given previously. If $W_1(\lambda_n - \varepsilon) < 0$, the sequence

$$W_1, W_2, \ldots, W_r \tag{5}$$

alternates in sign exactly $m - 2$ times. If $W_1(\lambda_n - \varepsilon) > 0$, sequence (5) alternates in sign exactly $m - 1$ times. These two conditions can be combined into the single condition, namely that the sequence

$$1, W_1, W_2, \ldots, W_r$$

alternates in sign exactly $m - 1$ times. By the classical theory of quadratic forms (see, for instance, [9]), the quadratic form corresponding to $W_r(\lambda_n - \varepsilon)$ must have the canonical representation (4).

It is easy to check the criterion for the classical choice $p_i = u_i$ $(i = 1, 2, \ldots, n-1)$. In this case, W_r is the product of the factors $(\lambda_j - \lambda)^{-1}$ $(j = 1, 2, \ldots, r = n-1)$. Let us also note that, for a given A, depending on the choice of $\{p_k\}$, the number r which appears in the criterion can be any value in the range $m - 1 \leq r \leq n - 1$, as can be seen by taking $p_i = u_i$ $(i = 1, 2, \ldots, r)$ and $p_{r+1} = p_{r+2} = \cdots = p_{n-1} = 0$.

In the case $2 \leq m = n$, a necessary and sufficient condition for (2) is that the quadratic form corresponding to $W_{m-1}(\lambda_n - \varepsilon)$ is negative-definite. When $2 \leq m < n$, this condition is sufficient for (2) for any choice of the vectors $p_m, p_{m+1}, \ldots, p_{n-1}$. This follows immediately from the previous section. However, this condition is not necessary for (2), as is shown by the following counterexample.

COUNTEREXAMPLE. Let $\lambda_1 < \lambda_2 = \lambda_3$, so that $m = 2$ and $n = 3$. Let $p_1 = u_2$ and $p_2 = u_1 + \alpha u_2$, where α is not zero but is otherwise arbitrary. Using p_1 alone, we obtain

$$W_1(\lambda_3 - \varepsilon) = 1/(\lambda_2 - \lambda_3 + \varepsilon) = 1/\varepsilon > 0,$$

for any $\varepsilon > 0$. For the moment, let us put $p_1' = p_2$. Then, we have

$$W_1(\lambda_3 - \varepsilon) = [1/(\lambda_1 - \lambda_3 + \varepsilon)] + \alpha^2/\varepsilon > 0$$

for any admissible $\varepsilon > 0$ which is sufficiently small. However, we see that

$$W_2(\lambda_3 - \varepsilon) = 1/\varepsilon(\lambda_1 - \lambda_3 + \varepsilon) < 0,$$

for any admissible and sufficiently small $\varepsilon > 0$. Therefore, we have

$\lambda(p_1) < \lambda_2\,(=\lambda_3)$ and $\lambda(p_2) < \lambda_2\,(=\lambda_3)$, but $\lambda(p_1, p_2) = \lambda_3$. This shows that to obtain equality (2) we need not have a negative-definite quadratic form corresponding to $W_{m-1}(\lambda_n - \varepsilon)$.

5. A Special Property of Operators Having Finite Traces

Let us add some remarks about the quantity $\lambda(p_1, p_2, \ldots, p_{n-1})$, which is basic for the new maximum–minumum theory. In the case of *a compact, negative (or positive) operator having a finite trace or at least a finite higher trace* which is computable, for instance in integral equations, Fichera [6] derived the following formula for $\lambda(p_1, p_2, \ldots, p_{n-1})$. Since $\lambda(p_1, p_2, \ldots, p_{n-1})$ is the first eigenvalue of $Q_r A Q_r$ and therefore has the largest absolute value, assuming for simplicity $\lambda_1^{(n-1)} < \lambda_2^{(n-1)}$, we can write, for sufficiently large k,

$$\mathbf{I}_1^{2k}(Q_r AQ_r) = \sum_{j=1}^{\infty} |\lambda_j^{(n-1)}|^{2k}$$

$$= |\lambda_1^{(n-1)}|^{2k}\left\{1 + \sum_{j=2}^{\infty} [|\lambda_j^{(n-1)}|/|\lambda_1^{(n-1)}|]^{2k}\right\}$$

so that

$$\lim_{k\to\infty} \{\mathbf{I}_1^{2k}(Q_r AQ_r)\}^{1/2k} = |\lambda_1^{(n-1)}|. \tag{1}$$

This formula is a representation of the value $\lambda_1^{(n-1)} = \lambda(p_1, p_2, \ldots, p_{n-1})$ in the case of positive, compact operators having finite traces, but it shows neither that $\lambda_1^{(n-1)} \leq \lambda_n$ for any choice of $p_1, p_2, \ldots, p_{n-1}$, nor that there exist $p_1, p_2, \ldots, p_{n-1}$ different from $u_1, u_2, \ldots, u_{n-1}$ for which equality holds.

Using Weinstein's new maximum–minimum theory we can answer the question of the existence of such p_i's for general operators of class $\mathscr{S}$.

6. The Existence of a Nonclassical Choice

In this section, we shall characterize all operators of class $\mathscr{S}$ for which the maximum of the minimum can be attained for a nonclassical choice of constraints. Usually, it suffices to consider $p_i = u_i$ as *the* classical choice, but in order to eliminate trivialities, we give the following more encompassing definition.

Definition. Let λ_n $(n > 1)$ be an eigenvalue of $A \in \mathscr{S}$ and let $p_1, p_2, \ldots, p_{n-1}$ be vectors in $\mathfrak{H}$. We say that the set $\{p_i\}$ is a *nonclassical choice* for the equality to hold in (3.1) if $\operatorname{sp}\{p_i\} \neq \operatorname{sp}\{u_1, u_2, \ldots, u_j\}$ for any j $(m - 1 \leq j \leq n - 1)$.

The following theorem is a generalization of remarks in [W18] and [W21].

Theorem 1. *For any operator $A \in \mathscr{S}$, there exists a nonclassical choice for every λ_n, except in the case when A is an operator on a finite-dimensional space and λ_n is its greatest eigenvalue.*

PROOF. We consider two cases.

CASE A. Let us first assume that there is a point in the spectrum of A, not necessarily an eigenvalue, that is greater than λ_n. Since λ_n is isolated and the spectrum is closed, we choose the smallest such element and call it ξ. We let $\mathfrak{V}$ be the finite-dimensional space $E_{\lambda_n}\mathfrak{H}$. In view of the existence of ξ, the space $\mathfrak{V}$ cannot be all of $\mathfrak{D}$ and therefore there exists a $w \in \mathfrak{V}^{\perp} \cap D$, $(w, w) = 1$ (see Lemma A.1). If $\lambda_n > \lambda_1$, we choose a real number α such that

$$0 < \alpha^2 \leq (\xi - \lambda_n)/(\lambda_n - \lambda_1)$$

and we put

$$p_1 = (1 + \alpha^2)^{-1/2}(u_1 + \alpha w)$$

and

$$p_2 = u_2, \ldots, \quad p_{n-1} = u_{n-1}.$$

In order to apply the criterion of Section 4, we compute the quantities $(R_\lambda p_i, p_k)$ $(i, k = 1, 2, \ldots, n-1)$ for $\lambda = \lambda_n - \varepsilon$, ε admissible. First of all, we have

$$\begin{aligned}(R_\lambda p_1, p_1) &= (1 + \alpha^2)^{-1}\left[(\lambda_1 - \lambda)^{-1} + \alpha^2 \int_{\xi - 0}^{\infty} (\mu - \lambda)^{-1}\, d(E_\mu w, w)\right] \\ &\leq (1 + \alpha^2)^{-1}\{(\lambda_1 - \lambda)^{-1} + [(\xi - \lambda_n)/(\xi - \lambda)(\lambda_n - \lambda_1)]\}.\end{aligned}$$

Putting $\lambda = \lambda_n - \varepsilon$, we have

$$\begin{aligned}(R_\lambda p_1, p_1) &\leq (1 + \alpha^2)^{-1}\{(\lambda_1 - \lambda_n + \varepsilon)^{-1} \\ &\qquad + [(\xi - \lambda_n)/(\xi - \lambda_n + \varepsilon)(\lambda_n - \lambda_1)]\} \\ &\leq (1 + \alpha^2)^{-1}[(\lambda_1 - \lambda_n + \varepsilon)^{-1} + (\lambda_n - \lambda_1)^{-1}] \\ &< 0.\end{aligned}$$

For $i = 2, 3, \ldots, n-1$, we have $(R_\lambda p_i, p_i) = 1/(\lambda_i - \lambda_n + \xi) < 0$ and for $i = m, m+1, \ldots, n-1$, we have

$$(R_\lambda p_i, p_i) = 1/\varepsilon > 0.$$

Obviously, the matrix $\{(R_\lambda p_i, p_k)\}$ is diagonal and the first $m-1$ diagonal elements are negative and the rest are positive. Therefore, by Theorem 4.1, we have indeed $\lambda(p_1, p_2, \ldots, p_{n-1}) = \lambda_n$ for a nonclassical choice $p_1, p_2, \ldots, p_{n-1}$. If $\lambda_1 = \lambda_n$, then we know by Lemma 1.1 that *any* choice yields the maximum of the minimum. Since the existence of ξ, shows that $\mathfrak{P}$ is not all of $\mathfrak{D}$, there is clearly a variety of nonclassical choices.

CASE B. We now consider the possibility that the eigenvalue λ_n is the largest point in the spectrum of A. Since λ_n and all preceding eigenvalues (if any) are isolated and have finite multiplicity, the operator A must be an operator on a finite-dimensional space. Moreover, we see that

$$\lambda_n = \max_{u \in \mathfrak{H}} R(u),$$

a statement which is, of course, trivial for finite-dimensional spaces. Suppose that the vectors $p_1, p_2, \ldots, p_{n-1} \in \mathfrak{H}$ are such that

$$\lambda(p_1, p_2, \ldots, p_{n-1}) = \lambda_n.$$

Let $\mathfrak{P} = \mathrm{sp}\{p_1, p_2, \ldots, p_{n-1}\}$ and let $\mathfrak{Q} = \mathfrak{P}^\perp$. Then, for any $w \in \mathfrak{Q}$, $w \neq 0$, we have

$$\lambda_n = \min_{u \in \mathfrak{Q}} R(u) \leq R(w) \leq \lambda_n = \max_{u \in \mathfrak{H}} R(u). \tag{1}$$

Therefore, we must have equality throughout (1) and w must satisfy the Euler equation

$$Aw = \lambda_n w.$$

Letting $\mathfrak{E}$ denote the eigenspace of λ_n, we see that $\mathfrak{Q} \subset \mathfrak{E}$, which implies that $\mathfrak{P} \supset \mathfrak{E}^\perp$, which in turn means that $\mathfrak{P}$ is a classical choice. Therefore, nonclassical choices, do not exist in this case, which proves the assertion.

7. A Stability Problem for Viscous Fluids

The theory of Navier–Stokes equations led Velte [V2] to numerous applications of the first type of intermediate problems. This matter is discussed here because it includes an illustration of the new maximum–minimum theory.

Omitting the hydrodynamic discussion, we formulate one of the problems considered (in two and three dimensions) by Velte. Among several domains studied by Velte and Sorger, we restrict our discussion to the case of the N-dimensional cube $\Omega = \{(x_1, x_2, \ldots, x_N) | 0 \leq x_k \leq \pi\}$ for any $N = 1, 2, \ldots$, in which case we give a more complete solution.

We denote by $D(u)$ the Dirichlet integral (omitting $dx_1\, dx_2 \cdots dx_N$)

$$D(u) = \int_\Omega \sum_{i=1}^{N} (\partial u/\partial x_i)^2$$

Let $v_1, v_2, \ldots, v_N$ be differentiable functions satisfying boundary conditions

$$v_i = 0 \quad \text{on} \quad \partial\Omega \qquad (i = 1, 2, \ldots, N) \tag{1}$$

and a divergence condition

$$\sum_{i=1}^{N} \partial v_i/\partial x_i = 0. \tag{2}$$

We want to find lower bounds for

$$\min\left\{[D(v_1) + D(v_2) + \cdots + D(v_N)] \Big/ \sum_{i=1}^{N} (v_i, v_i)\right\} \tag{3}$$

under conditions (1) and (2). Obviously, we assume that not all $v_i = 0$. By omitting condition (2), we obtain

$$\min\left[\sum_{i=1}^{N} D(v_i) \Big/ \sum_{i=1}^{N} (v_i, v_i)\right], \tag{4}$$

$$v_i = 0 \quad \text{on} \quad \partial\Omega,$$

which is not greater than the more restricted minimum.

We now derive a basic lemma.

Lemma 1. *Let $\mathfrak{B}$ be any subspace of differentiable functions on Ω for which the minima in* (5) *exist. Then,*

$$\min_{\substack{v \in \mathfrak{B} \\ v \neq 0}} [D(v)/(v, v)] = \min_{v_i \in \mathfrak{B}} \left[\sum_{i=1}^{N} D(v_i) \Big/ \sum_{i=1}^{N} (v_i, v_i)\right]. \tag{5}$$

PROOF. Noting that not all v_i can be zero, we assume that $v_j \in \mathfrak{B}$, $v_j \neq 0$, is such that

$$D(v_j)/(v_j, v_j) \leq D(v_i)/(v_i, v_i)$$

for all $v_i \in \mathfrak{B}$, $v_i \neq 0$. Then, we have

$$\begin{aligned}
[D(v_j)/(v_j, v_j)] \sum_{i=1}^{N} (v_i, v_i) &= \sum_{i=1}^{N} [D(v_j)/(v_j, v_j)](v_i, v_i) \\
&\leq \sum_{\substack{i=1 \\ (v_i, v_i) \neq 0}}^{N} [D(v_i)/(v_i, v_i)](v_i, v_i) \\
&= \sum_{i=1}^{N} D(v_i),
\end{aligned}$$

so that

$$\min_{\substack{v \in \mathfrak{V} \\ v \neq 0}}[D(v)/(v, v)] \leq \min_{\substack{v_i \in \mathfrak{V} \\ i=1,2,\ldots,N}} \left[\sum_{i=1}^{N} D(v_i) \Big/ \sum_{i=1}^{N} (v_i, v_i)\right]. \tag{6}$$

Since we can put $N - 1$ of the v_i's equal to zero, it always is the case that

$$\min_{\substack{v_i \in \mathfrak{V} \\ i=1,2,\ldots,N}} \left[\sum_{i=1}^{N} D(v_i) \Big/ \sum_{i=1}^{N} (v_i, v_i)\right] \leq D(v)/(v, v),$$

for any nonzero $v \in \mathfrak{V}$. Therefore, we must have equality in (6).

In view of this lemma, we can conclude that

$$\min\,[D(v)/(v, v)] \qquad (v = 0 \quad \text{on} \quad \partial\Omega), \tag{7}$$

is not greater than (4). The Euler equation corresponding to (7) is

$$\Delta v + \mu v = 0 \quad \text{in} \quad \Omega \qquad (v = 0 \quad \text{on} \quad \partial\Omega). \tag{8}$$

As is well known, the operator in (8) is of class $\mathscr{S}$ and has known spectrum. This will be used as our base problem. To obtain better lower bounds, we observe, with Velte, that condition (2) implies that for every differentiable function f we have

$$\int_\Omega f\left[\sum_{i=1}^{n} (\partial v_i/\partial x_i)\right] = 0,$$

which, because of condition (1) and Green's identity, yields a new condition,

$$\int_\Omega \sum_{i=1}^{N} v_i\left(\frac{\partial f}{\partial x_i}\right) = 0. \tag{9}$$

Therefore, the minimum in (4) under condition (9) is not greater than the corresponding minimum under condition (2). Now, instead of using all differentiable functions f in (9), we only use N functions, namely $f_1 = x_1, f_2 = x_2, \ldots, f_N = x_N$. Noting that $\partial f_i/\partial x_i = 1$, we see that the minimum in (4), under the conditions (mean values equal to zero)

$$\int_\Omega v_i = 0 \qquad (i = 1, 2, \ldots, N), \tag{10}$$

is not greater than the same minimum under (9). By Lemma 1, we know that

$$\min[D(v)/(v, v)] \qquad \left(v = 0 \quad \text{in} \quad \partial\Omega; \quad \int_\Omega v = 0\right) \tag{11}$$

is equal to the minimum in (4) under (10). Putting $p_i = \partial f_i/\partial x_i = 1$, we see that *this problem is actually an intermediate problem*. It can be solved in terms of the base problem (8) and yields improved lower bounds.

The Euler equation of the intermediate problem (11) is

$$\Delta v + \mu v = \alpha \qquad \left(v = 0 \quad \text{on} \quad \partial\Omega; \quad \int_\Omega v = 0; \quad \alpha \quad \text{a scalar}\right).$$

The eigenvalues of the base problem (8) are given by

$$\lambda^{(0)} = \sum_{k=1}^{N} \beta_k{}^2,$$

where the β_k take on all positive integer values, so that

$$\lambda_1^{(0)} = N, \quad \lambda_2^{(0)} = N + 3, \ldots,$$

and the corresponding eigenfunctions are given by

$$u^{(0)} = (2/\pi)^{N/2}(\sin \beta_1 x_1)(\sin \beta_2 x_2) \cdots \sin \beta_N x_N .$$

We know by (3.1.2) that the minimum in (11) cannot exceed $\lambda_2^{(0)} = N + 3$. In order to determine whether the minimum is equal to $N + 3$, we use the results of the *new maximum–minimum theory*. In this case, we have to evaluate $W(\lambda_2^{(0)})$, which turns out to be

$$\begin{aligned} W(\lambda_2^{(0)}) &= W(N + 3) \\ &= (8/\pi)^N \sum_{h_1, h_2, \ldots, h_N} \{(2h_1 + 1)^2 \cdots (2h_N + 1)^2 \\ &\qquad \times [(2h_1 + 1)^2 + \cdots + (2h_N + 1)^2 - N - 3]\}^{-1}. \end{aligned}$$

The computations, which were performed by J. K. Oddson using Bernoulli numbers (see [W21]), show that

$$W(N + 3) < 0 \qquad (N = 1, 2, \ldots, 9)$$

and

$$W(N + 3) > 0 \qquad (N = 10, 11, \ldots).$$

This gives the completely unexpected result that the minimum in (11) is exactly equal to $\lambda_2^{(0)} = N + 3$ for dimensions $N = 1, 2, \ldots, 9$ and is strictly less than $\lambda_2^{(0)}$ for all other dimensions $N = 10, 11, \ldots$. Incidentally, the case $N = 1$ was already discussed in [W9] and is probably the oldest example of a nonclassical choice in the maximum–minimum theory.

In this section, we have considered unbounded differential operators.

We have again derived the results directly from the differential problem. However, let us note that we could have introduced the inverse compact Green's operator G here and proceeded in a way nearly identical to that given in Section 4.12.

8. Some Inequalities for Higher Eigenvalues

As we have seen in Chapters 3 and 4, we have the following chain of inequalities

$$\lambda_i^{(0)} \le \lambda_i^{(n)} \le \lambda_{n+i}^{(0)} \qquad (i = 1, 2, \ldots). \tag{1}$$

The left side ($\lambda_i^{(0)} \le \lambda_i^{(n)}$) follows directly from the principle of monotonicity (2.5.3), and the right side ($\lambda_i^{(n)} \le \lambda_{n+i}^{(0)}$) from a recursive application of Weyl's inequality (see Section 3.1) or from the new maximum–minimum theory for compact operators. These inequalities will play a part in Sections 8.2 and 8.6 as well as in the perturbation problems of Section 9.2.

In the present section, we shall discuss the equality signs in all of the inequalities in (1), extending the results obtained for the inequality

$$\lambda_1^{(n-1)} \le \lambda_n. \tag{2}$$

In Section 4, we saw that a necessary condition for equality in (2) was that the dimension r of $\mathfrak{P} = \text{sp}\{p_1, p_2, \ldots, p_{n-1}\}$ had to satisfy an inequality $m - 1 \le r$. For the new maximum–minimum theory to be self-contained, the proof of this fact was independent of Weyl's lemma. Here, our approach is slightly different. By using Weyl's lemma in the following, we first give necessary conditions for the equalities to hold in (1) and then develop necessary and sufficient conditions; see [S 11].

Theorem 1. *A necessary condition for the equality*

$$\lambda_i^{(n)} = \lambda_{n+i}^{(0)} \tag{3}$$

to hold is that the dimension r of $\text{sp}\{p_1, p_2, \ldots, p_n\}$ *satisfy the inequality*

$$m - i \le r, \tag{4}$$

where

$$m = \min\{j \mid \lambda_j = \lambda_{n+i}^{(0)}\}.$$

Proof. Letting $\{p_1', p_2', \ldots, p_r'\}$ be any linearly independent subset of $\{p_1, p_2, \ldots, p_n\}$, we have from Weyl's inequalities (3.1.2)

$$\lambda_i^{(n)} = \lambda_i^{(r)} \le \lambda_{r+i}^{(0)} \le \lambda_{n+i}^{(0)}. \tag{5}$$

In order that $\lambda_i^{(n)} = \lambda_{n+i}^{(0)}$, we must have equality throughout (5). In particular, we require that

$$\lambda_{r+i}^{(0)} = \lambda_{n+i}^{(0)},$$

and therefore we have a necessary condition (4).

Let us remark that, if $\mathfrak{H}$ is infinite-dimensional, the other equality,

$$\lambda_i^{(0)} = \lambda_i^{(n)}, \tag{6}$$

does not require an inequality for the dimension of sp$\{p_1, p_2, \ldots, p_n\}$. In fact, by choosing *any* number of p_i's that are orthogonal to $u_1, u_2, \ldots, u_i$, we see from Section 4.4 that the eigenvalues $\lambda_1^{(0)}, \lambda_2^{(0)}, \ldots, \lambda_i^{(0)}$ are all persistent, and therefore equality (6) clearly holds.

In the following, we assume that $p_1, p_2, \ldots, p_n$ are independent. Otherwise, we would have to make some slight modifications. Let us now formulate our criteria.

Theorem 2. *For an arbitrary choice of linearly independent vectors $p_1, p_2, \ldots, p_n \in \mathfrak{H}$, we have the inequalities*

$$\lambda_i^{(n)} \leq \lambda_{n+i}^{(0)} \qquad (i = 1, 2, \ldots). \tag{7}$$

For a given index i, we have the following necessary and sufficient condition for the equality:

$$\lambda_i^{(n)} = \lambda_{n+i}^{(0)} \tag{8}$$

to hold: For any admissible ε, the quadratic form defined by the symmetric matrix

$$\{(R_\lambda p_j, p_k)\} \qquad (j, k = 1, 2, \ldots, n) \tag{9}$$

evaluated at $\lambda = \lambda_{n+i}^{(0)} - \varepsilon$ has in canonical coordinates the diagonal form

$$-\mathbf{x}_1{}^2 - \mathbf{x}_2{}^2 - \cdots - \mathbf{x}_{m-i}^2 \pm \mathbf{x}_{m-i+1}^2 \pm \cdots \pm \mathbf{x}_{m-1}^2 + \mathbf{x}_m{}^2 + \cdots + \mathbf{x}_n{}^2, \tag{10}$$

where $m = \min\{j \mid \lambda_j^{(0)} = \lambda_{n+i}^{(0)}\}$.

PROOF. We first note that the maxi-mini-max principle (Section 3.6) yields a set of inequalities, namely

$$\lambda_i^{(n)} \leq \lambda_{i+1}^{(n-1)} \leq \cdots \leq \lambda_{n-1+i}^{(1)} \leq \lambda_{n+i}^{(0)}. \tag{11}$$

Now, let

$$m_h = \min\{j \mid \lambda_j^{(h)} = \lambda_{n+i}^{(0)}\} \qquad (h = 1, 2, \ldots, n).$$

We assume for the moment that

$$(p_j, p_k) = \delta_{jk} \qquad (j, k = 1, 2, \ldots, n). \tag{12}$$

Now, let s_n denote the number of times that

$$m_h = m_{h-1} - 1 \tag{13}$$

holds in going from the base problem to the nth intermediate problem. Using Eq. (13) n times, we see that s_n is equal to the number of negative factors in the product

$$W_{01} W_{12} \cdots W_{n-1,n}$$

evaluated at the point $\lambda_{n+1}^{(0)} - \varepsilon$. As in Theorem 4.1, we see that s_n is equal to the number of times the sequence $1, W_1, W_2, \ldots, W_n$, evaluated at $\lambda_{n+i}^{(0)} - \varepsilon$, alternates in sign. Therefore, by the classical theory of quadratic forms, s_n is the number of negative terms in the diagonal form corresponding to $W_n(\lambda_{n+i}^{(0)} - \varepsilon)$. As in Section 4, a change of the basis of $\mathfrak{P}$ corresponds to a change of the basis of the quadratic form, so that the diagonal form depends only on the subspace $\mathfrak{P}$ and not on the basis chosen. Therefore, the final result does not require assumption (12).

To prove the necessity of (10), let us suppose that $\lambda_i^{(n)} = \lambda_{n+i}^{(0)}$. Then, it follows that $1 \le m - s_n = m_n \le i$. Therefore, we have $m - i \le s_n \le m - 1$, which means that the quadratic form corresponding to (9) has the canonical representation (10). On the other hand, to prove the sufficiency of (10), let us suppose that we have (10) but that $\lambda_i^{(n)} < \lambda_{n+i}^{(0)}$. This means that

$$s_n \ge m - i \tag{14}$$

and that there is an index h $(0 \le h < n)$ for which

$$\lambda_{n+i-h-1}^{(h+1)} < \lambda_{n+i-h}^{(h)} = \cdots = \lambda_{n+i}^{(0)}.$$

Therefore, we have

$$m_h = n + i - h \tag{15}$$

and

$$W_{h,h+1}(\lambda_{n+i}^{(0)} - \varepsilon) > 0. \tag{16}$$

Equation (15) implies that

$$s_h = m - m_h = m + h - i - n.$$

However, inequality (16) means that Eq. (13) holds at most $n - 1 - h$ times from the hth intermediate problem to the nth intermediate problem. Therefore, we have

$$s_n \le s_h + n - 1 - h = m + h - i - n + n - 1 - h = m - i - 1,$$

which contradicts assumption (10).

Theorem 3. *For an arbitrary choice of linearly independent vectors* $p_1, p_2, \ldots, p_n \in \mathfrak{H}$, *we have the inequalities*

$$\lambda_i^{(0)} \leq \lambda_i^{(n)} \qquad (i = 1, 2, \ldots). \tag{17}$$

For a given index i, we have the following necessary and sufficient condition for the equality

$$\lambda_i^{(0)} = \lambda_i^{(n)} \tag{18}$$

to hold: For any admissible ε, *the quadratic form defined by the symmetric matrix*

$$\{(R_\lambda p_j, p_k)\} \qquad (j, k = 1, 2, \ldots, n) \tag{19}$$

evaluated at $\lambda = \lambda_i^{(0)} + \varepsilon$ *has in canonical coordinates the diagonal from*

$$\pm \mathbf{x}_1{}^2 \pm \mathbf{x}_2{}^2 \pm \cdots \pm \mathbf{x}_{M-i}^2 + \mathbf{x}_{M-i+1}^2 + \cdots + \mathbf{x}_n{}^2, \tag{20}$$

where $M = \max\{j \mid \lambda_j^{(0)} = \lambda_i^{(0)}\}$.

Note that it follows immediately from (5) that, if $n \leq m - i$, then equality (18) holds for *any* choice of $p_1, p_2, \ldots, p_n$.

Proof. The inequalities follow from the maximum–minimum or minimum–maximum principles. In proving our necessary and sufficient conditions for equality (18), we temporarily assume as above that the p_j satisfy (12). If we let M_h denote the largest index in the hth intermediate problem such that $\lambda_i^{(0)} = \lambda_{M_h}$, we see from the table in Section 2 that either $M_{h+1} = M_h$ or

$$M_{h+1} = M_h - 1, \tag{21}$$

for $h = 1, 2, \ldots, n - 1$. Now, let t_n denote the number of times Eq. (21) holds in going from the base problem to the nth intermediate problem. From the table and the monotonicity of $W_{jk}(\lambda)$, it is clear that t_n is the number of negative factors in the product

$$W_{01} W_{12} \cdots W_{n-1,n}$$

evaluated at the point $\lambda_i^{(0)} + \varepsilon$. Proceeding as in the previous proofs, we see that t_n is the number of negative terms in the diagonal form of (20). Our necessary and sufficient condition for equality (18) now follows since $\lambda_i^{(0)} = \lambda_i^{(n)}$ if and only if $i \leq M_n$. But we have by (21) that $M_n = M - t_n$. Therefore, equality (18) holds if and only if $t_n \leq M - i$, which proves the theorem.

Again, as in Section 6, these theorems enable one to *predict* nonclassical choices of vectors, $p_1, p_2, \ldots, p_n$ that yield equality signs. It is remarkable that, in all of these theorems connected with the maximum–minimum theory, the inequalities hold for all choices of $p_1, p_2, \ldots, p_n$, while already in the classical sufficient condition for equality to hold in (2), the eigenelements $u_1, u_2, \ldots, u_n$ of the operator are explicitly used. It would therefore be surprising if, in the formulation of necessary and sufficient conditions, none of the eigenelements had to be used.

Chapter Eight

Inequalities for Eigenvalues of Parts and Projections of Operators

1. Some Preliminary Lemmas

In this chapter we shall discuss some inequalities due to several authors. These inequalities relate the eigenvalues of a given operator to the eigenvalues of its projections onto certain closed subspaces (see Section 3.2). For several of the inequalities, we give criteria for equality to hold, paralleling Weinstein's new maximum–minimum theory.

We start with several lemmas which will be used subsequently, the first of which is a modification of a lemma in the book by Hamburger and Grimshaw [16, p. 76].

Lemma 1. *Let α and γ be any nonnegative constants and let β be any complex number such that*

$$f(z) = \alpha z\bar{z} + \beta z + \bar{\beta}\bar{z} + \gamma \geq 0$$

for all complex z. Then, we have

$$\beta\bar{\beta} \leq \alpha\gamma \tag{1}$$

and

$$f(z) \leq (1 + z\bar{z})(\alpha + \gamma), \tag{2}$$

with equality in (2) if and only if there is some z_0 such that $f(z_0) = 0$.

PROOF. We write $z = re^{i\theta}$, $\beta = \rho e^{i\phi}$, where r and ρ are nonnegative and θ and ϕ are real. Then, we have

$$f(z) = \alpha r^2 + 2\rho r\cos(\theta + \phi) + \gamma \geq 0,$$

and therefore,

$$f(z) \geq \alpha r^2 - 2\rho r + \gamma \geq 0. \tag{3}$$

If $\alpha = 0$ or $\gamma = 0$, we see from (3) that ρ must be zero, and inequalities (1) and (2) follow trivially. Henceforth, we assume α, $\gamma > 0$. Then, it follows that

$$f(z) = \alpha r^2 - 2\rho r + \gamma \geq 0$$

only if $\cos(\theta + \phi) = -1$. Moreover, the inequality on the right implies that the discriminant $4\rho^2 - 4\alpha\gamma$ is nonpositive, so that inequality (1) holds and $\alpha\gamma = \beta\bar{\beta}$, if and only if there exists a z_0 such that $f(z_0) = 0$. Since $\rho \leq (\alpha\gamma)^{1/2}$, we can write

$$\rho r \leq r(\alpha\gamma)^{1/2}$$

for all r. Since the geometric mean is dominated by the arithmetic mean, we have

$$\rho r \leq r(\alpha\gamma)^{1/2} \leq \tfrac{1}{2}(\alpha + \gamma r^2),$$

so that we obtain

$$\begin{aligned} f(z) &= \alpha r^2 + 2\rho r \cos(\theta + \phi) + \gamma \\ &\leq \alpha r^2 + 2\rho r + \gamma \\ &\leq \alpha r^2 + \alpha + \gamma r^2 + \gamma. \end{aligned}$$

which yields the desired inequality (2).

Lemma 2. *Let $f(z)$ be as in Lemma 1. If there exists a complex number z_0 such that $f(z_0) = (1 + z_0 \bar{z}_0)(\alpha + \gamma)$, then $\alpha\gamma = \beta\bar{\beta}$.*

PROOF. We must consider four cases.

CASE A. If $\alpha = 0$, we have

$$f(z) = 2\,\mathbf{Re}(\beta z) + \gamma \geq 0$$

for all complex z, which is only possible for $\beta = 0$. Therefore, $\alpha\gamma = \beta\bar{\beta} = 0$, so the result follows trivially.

CASE B. If $\gamma = 0$, we have

$$f(z) = \alpha z\bar{z} + \beta z + \bar{\beta}\bar{z} \geq 0 \tag{4}$$

for all complex z. Therefore, for every complex $z \neq 0$, we obtain

$$\alpha + (\beta/\bar{z}) + (\bar{\beta}/z) \geq 0.$$

Putting $\zeta = 1/\bar{z}$, we get

$$\alpha + 2\,\mathbf{Re}(\beta\zeta) \geq 0$$

for all complex ζ, which is essentially case A.

CASE C. Suppose that

$$f(z_0) = (1 + z_0\,\bar{z}_0)(\alpha + \gamma)$$

for $z_0 = 0$. Then, it follows that

$$f(z_0) = \gamma = (1 + 0)(\alpha + \gamma)$$

which implies that $\alpha = 0$, and we again have case A.

CASE D. Suppose that there exists $z_0 \neq 0$ such that

$$\alpha z_0\,\bar{z}_0 + \beta z_0 + \bar{\beta}\bar{z}_0 + \gamma = (1 + z_0\,\bar{z}_0)(\alpha + \gamma). \tag{5}$$

Canceling like terms, we can write Eq. (5) as

$$\alpha - \beta z_0 - \bar{\beta}\bar{z}_0 + \gamma z_0\,\bar{z}_0 = 0. \tag{6}$$

We now put $\zeta_0 = -1/\bar{z}_0$ in (6) and multiply by $|\zeta_0|^2$ in order to obtain

$$\alpha\zeta_0\bar{\zeta}_0 + \beta\zeta_0 + \bar{\beta}\bar{\zeta}_0 + \gamma = 0. \tag{7}$$

Applying Lemma 1 to Eq. (7), we get $\alpha\gamma = \beta\bar{\beta}$, which concludes the proof of the lemma.

For the sake of completeness, we give the following criterion for the equality $\alpha\gamma = \beta\bar{\beta}$.

Lemma 3. *Let $f(z)$ be as in Lemma 1. Then, $\alpha\gamma = \beta\bar{\beta}$ if and only if either*

$$f(0) = 0 \tag{8}$$

or there exists a ζ_0 such that

$$f(\zeta_0) = (1 + \zeta_0\bar{\zeta}_0)(\alpha + \gamma). \tag{9}$$

PROOF. The sufficiency of (8) follows from Lemma 1 and that of (9) from Lemma 2. For the necessity, suppose that $\alpha\gamma = \beta\bar{\beta}$. Then, by Lemma 1, there exists z_0 such that $f(z_0) = 0$. If $z_0 = 0$, we have (8). If $z_0 \neq 0$, we put $\zeta_0 = -1/\bar{z}_0$. Then, we have

$$0 = \alpha z_0\,\bar{z}_0 + \beta z_0 + \bar{\beta}\bar{z}_0 + \gamma = z_0\,\bar{z}_0\{\alpha - \beta\zeta_0 - \overline{\beta\zeta_0} + \gamma\zeta_0\bar{\zeta}_0\}.$$

Therefore, we get

$$\alpha - \beta\zeta_0 - \overline{\beta\zeta_0} + \gamma\zeta_0\bar{\zeta}_0 = 0. \tag{10}$$

Adding $f(\zeta_0) = \alpha\zeta_0\bar{\zeta}_0 + \beta\zeta_0 + \overline{\beta\zeta_0} + \gamma$ to both sides of Eq. (10), we obtain (9), which concludes the proof.

2. Aronszajn's Inequality

In connection with his investigation of the first type of intermediate problems, Aronszajn found a remarkable inequality. In this section, we discuss this inequality for operators in $\mathscr{S}$ that are not only bounded below but are *also bounded above*. In the case of a negative-definite, compact operator, we shall give a criterion for the equality sign to hold.

Let $\mathfrak{H}'$ be a closed subspace of $\mathfrak{H}$ and let $\mathfrak{H}'' = \mathfrak{H} \ominus \mathfrak{H}'$. Denote by P' and P'' the projection operators onto $\mathfrak{H}'$ and $\mathfrak{H}''$, respectively.

For the following, we assume that A, $P'AP'$, and $P''AP''$ are in class $\mathscr{S}$ and that the indices of the eigenvalues are meaningful. These assumptions are certainly satisfied for compact operators.

Let $\lambda_1 \leq \lambda_2 \leq \cdots$, $\lambda_1' \leq \lambda_2' \leq \cdots$, and $\lambda_1'' \leq \lambda_2'' \leq \cdots$ be the eigenvalues, and $u_1, u_2, \ldots, u_1', u_2', \ldots$, and $u_1'', u_2'', \ldots$, be the corresponding eigenvectors of the operators A, $P'AP'$, and $P''AP''$, respectively. Let $\mu = \sup\{(Au, u)/(u, u)\}$ for $u \neq 0$, $u \in \mathfrak{H}$. Note that, since A is bounded, its domain is all of $\mathfrak{H}$.

Theorem 1. *The eigenvalues of the operators A, $P'AP'$, and $P''AP''$ satisfy the inequalities*

$$\lambda_i' + \lambda_j'' \leq \lambda_{i+j-1} + \mu \qquad (i, j = 1, 2, \ldots). \tag{1}$$

PROOF. For any $u \in \mathfrak{H}$, $(u, u) = 1$, we can write $u = \sigma v + \tau w$, where $v \in \mathfrak{H}'$, $w \in \mathfrak{H}''$, $(v, v) = (w, w) = 1$, and $|\sigma|^2 + |\tau|^2 = 1$. If $\tau = 0$, we have $R(u) = R(v)$ and $R(w) \leq \mu$ for any $w \in \mathfrak{H}''$. Therefore, we obtain

$$R(v) + R(w) \leq R(u) + \mu. \tag{2}$$

Let us now prove that (2) is true for $\tau \neq 0$. By direct computation, we have

$$\begin{aligned} 0 \leq \mu - R(u) = \sigma\bar{\sigma}[\mu - R(v)] &- \sigma\bar{\tau}([A - \mu I]v, w) \\ &- \overline{\bar{\sigma}\tau([A - \mu I]v, w)} + \tau\bar{\tau}[\mu - R(w)]. \end{aligned}$$

Dividing by $|\tau|^2$, we can write

$$0 \le [\mu - R(u)]/|\tau|^2 = |\sigma/\tau|^2[\mu - R(v)] - (\sigma/\tau)([A - \mu I]v, w) - (\bar{\sigma}/\bar{\tau})([A - \mu I]v, w) + [\mu - R(w)].$$

Let us now apply Lemma 1.1.

We set $z = \sigma/\tau$, $\alpha = \mu - R(v)$, $\beta = -([A - \mu I]v, w)$, and $\gamma = \mu - R(w)$. Applying inequality (1.2), we obtain the inequality

$$[\mu - R(u)]/|\tau|^2 \le \{1 + |\sigma/\tau|^2\}\{2\mu - R(v) - R(w)\}.$$

Multiplying by $|\tau|^2$, we again get inequality (2). Therefore (2) is true for every nonzero $u \in \mathfrak{H}$. Now, let $\mathfrak{U}'_{i-1} = \text{sp}\{u_1', u_2', \ldots, u'_{i-1}\}$ and $\mathfrak{U}''_{j-1} = \text{sp}\{u_1'', u_2'', \ldots, u''_{j-1}\}$ for $i, j = 2, 3, \ldots$ and set $\mathfrak{U}_0' = \mathfrak{U}_0'' = \{0\}$. Using inequality (2) and the *maximum–minimum principle*, we get

$$\begin{aligned}\lambda_i' + \lambda_j'' &= \min_{\substack{v \perp \mathfrak{A}'_{i-1} \\ v \in \mathfrak{H}'}} R(v) + \min_{\substack{w \perp \mathfrak{A}''_{j-1} \\ w \in \mathfrak{H}''}} R(w) \\ &\le \min_{\substack{u \perp \mathfrak{A}'_{j-1} \oplus \mathfrak{A}''_{i-1} \\ u \in \mathfrak{H}}} R(u) + \mu \le \lambda_{i+j-1} + \mu,\end{aligned}$$

which proves Aronszajn's inequality.

Corollary 1. *If A is nonpositive, then we have the simpler inequality*

$$\lambda_i' + \lambda_j'' \le \lambda_{i+j-1} \qquad (i, j = 1, 2, \ldots). \tag{3}$$

Assuming from now on that A is compact and strictly negative definite, we shall establish necessary and sufficient conditions for the equality sign to hold in inequality (3). See Stenger [S11]. In order to do this, we first establish necessary conditions.

Lemma 1. *For a given pair of indices i, j, a necessary condition for the equality*

$$\lambda_i' + \lambda_j'' = \lambda_{i+j-1} \tag{4}$$

to be satisfied in inequality (3) *is that either*

$$\lambda_i' = 0 \qquad \text{or} \qquad \lambda_j'' = 0.$$

PROOF. In view of the proof of Theorem 1, equality (4) can be satisfied only if there exists a vector $u_0 = \sigma_0 v_0 + \tau_0 w_0$ such that $\lambda_i' = R(v_0)$. $\lambda_j'' = R(w_0)$, and $\lambda_{i+j-1} = R(u_0)$. If $\tau_0 = 0$, we have $\lambda_i' = \lambda_{i+j-1}$ and

$\lambda_j'' = 0$. If $\sigma_0 = 0$, we have $\lambda_j'' = \lambda_{i+j-1}$ and $\lambda_i' = 0$. Suppose that both $\sigma_0 \neq 0$ and $\tau_0 \neq 0$. In this case, we obtain

$$f(\sigma_0/\tau_0) = -R(u_0)/|\tau_0|^2 = \{1 + |\sigma_0/\tau_0|^2\}[-R(v_0) - R(w_0)],$$

where now $z_0 = \sigma_0/\tau_0$, $\alpha = -R(v_0)$, $\beta = -(Av_0, w_0)$, and $\gamma = -R(w_0)$ in the notation of Section 1. It follows from Lemma 1.2 that

$$R(v_0)R(w_0) = |(Av_0, w_0)|^2. \tag{5}$$

Now, let

$$v_0 = \sum_{k=1}^{\infty} v_k u_k \qquad \text{and} \qquad w_0 = \sum_{k=1}^{\infty} \omega_k u_k.$$

Then,

$$-R(v_0) = \sum_{k=1}^{\infty} -\lambda_k |v_k|^2, \qquad -R(w_0) = \sum_{k=1}^{\infty} -\lambda_k |\omega_k|^2,$$

$$-(Av_0, w_0) = \sum_{k=1}^{\infty} -\lambda_k v_k \bar{\omega}_k.$$

Let

$$\xi_k = (-\lambda_k)^{1/2} v_k \qquad \text{and} \qquad \eta_k = (-\lambda_k)^{1/2} \omega_k \qquad (k = 1, 2, \ldots).$$

Substituting these quantities in Eq. (5), we obtain

$$\left(\sum_{k=1}^{\infty} |\xi_k|^2\right)\left(\sum_{k=1}^{\infty} |\eta_k|^2\right) = \left|\sum_{k=1}^{\infty} \xi_k \bar{\eta}_k\right|^2.$$

This means that there is a constant ρ such that $\xi_k = \rho\eta_k$ $(k = 1, 2, \ldots)$. Since A is negative-definite, we have $\lambda_k \neq 0$ $(k = 1, 2, \ldots)$, so that $v_k = \rho\omega_k$ $(k = 1, 2, \ldots)$. Therefore, we get

$$0 = (v_0, w_0) = \sum_{k=1}^{\infty} v_k \bar{\omega}_k = \rho \sum_{k=1}^{\infty} |\omega_k|^2 = \rho,$$

so that $v_0 = 0$, which contradicts the assumption that $(v_0, v_0) = 1$ and completes the proof.

Lemma 2. *A necessary condition for equality* (4) *to hold is that either* $\dim \mathfrak{H}' \leq i - 1$ *or* $\dim \mathfrak{H}'' \leq j - 1$.

Proof. From Lemma 1, we see that either $\lambda_i' = 0$ or $\lambda_j'' = 0$. By symmetry, it suffices to consider only one case, so let us suppose that $\lambda_i' = 0$. Assume that $\dim \mathfrak{H}' \geq i$ and let $\mathfrak{B}$ be any i-dimensional subspace of $\mathfrak{H}'$.

At this point, we use the *minimum–maximum principle*, which is not used in the proof of inequality (3), to obtain

$$0 = \lambda_i' \leq \max_{u \in \mathfrak{B}} \frac{(P'AP'u, u)}{(u, u)} = \max_{u \in \mathfrak{B}} \frac{(Au, u)}{(u, u)} \leq 0. \tag{6}$$

Therefore, we must have equality throughout (6). In particular, this means that there exists a $u_0 \in \mathfrak{B}$, $(u_0, u_0) = 1$, such that $(Au_0, u_0) = 0$, which contradicts the assumption that A is negative-definite.

We can now apply the *new maximum–minimum theory* to give the following criterion for equality (4), see [S11].

Theorem 2. *A necessary condition for the equality* (4),

$$\lambda_i' + \lambda_j'' = \lambda_{i+j-1},$$

to be satisfied is that the dimension r of the subspace $\mathfrak{H}''$ (*or* $\mathfrak{H}'$) *satisfy the inequalities*

$$m - i \leq r \leq j - 1 \tag{7}$$

where $m = \min\{h \,|\, \lambda_h = \lambda_{i+j-1}\}$. *Let* $p_1, p_2, \ldots, p_r$ *be a basis for* $\mathfrak{H}''$. *If inequality* (7) *is satisfied, a necessary and sufficient condition for equality* (4) *to be satisfied is that, for any admissible* $\varepsilon > 0$, *the quadratic form with the symmetric matrix*

$$\{(R_\lambda p_h, p_k)\} \qquad (h, k = 1, 2, \ldots, r)$$

evaluated at $\lambda_{i+j-1} - \varepsilon$ *has in canonical coordinates the diagonal form*

$$-\mathbf{x}_1^2 - \mathbf{x}_2^2 \cdots - \mathbf{x}_{m-i}^2 \pm \mathbf{x}_{m-i+1}^2 \pm \cdots \pm \mathbf{x}_{m-1}^2 + \mathbf{x}_m^2 + \cdots + \mathbf{x}_r^2.$$

Proof. In view of Lemma 2, we have the necessary condition $r \leq j - 1$. If $r \leq j - 1$, we see that $\lambda_i' = \lambda_i^{(j-1)}$, where $\mathfrak{H}''$ plays the role of the subspace $\mathfrak{P} = \mathrm{sp}\{p_k\}$ in Theorem 7.8.1. It follows from Theorem 7.8.1 that a necessary condition for equality (4), which is now reduced to

$$\lambda_i^{(j-1)} = \lambda_{i+j-1},$$

is that $m - i \leq r$. The necessary and sufficient condition stated above now follows immediately from Theorem 7.8.1.

At the present time, the criterion for the more general inequalities (1) is still an open question.

3. An Estimate of the Rate of Convergence for Intermediate Problems

As we have mentioned, it is usually necessary to have upper and lower bounds for a given eigenvalue in order to estimate the error in the computed approximation. With regard to the question of the rate of convergence in the first type of intermediate problems, Weinberger [W3] obtained a theoretical result by using inequality (2.3) and a special generating sequence in the given subspace $\mathfrak{P}$.

Consider a given problem of the first type (Section 4.2)

$$QAu = \lambda u, \qquad Pu = 0,$$

where A is here restricted to be a compact, negative operator. Let us put

$$p_i = Pu_i \qquad (i = 1, 2, \ldots).$$

It is easy to see that the sequence $\{p_i\}$ is complete in $\mathfrak{P}$. In fact, if $p \in \mathfrak{P}$ satisfies the orthogonality conditions

$$(p, p_i) = 0 \qquad (i = 1, 2, \ldots),$$

then we have

$$(p, Pu_i) = (p, u_i) = 0 \qquad (i = 1, 2, \ldots),$$

which means that $p = 0$.

We shall now show that eigenvalues of the intermediate problems obtained from this sequence satisfy the inequality

$$\lambda_k^{(\infty)} - \lambda_k^{(n)} \leq |\lambda_{n+1}^{(0)}|, \tag{1}$$

which provides an estimate of the rate of convergence.

First of all, we put $\mathfrak{P}_n = \text{sp}\{p_1, p_2, \ldots, p_n\}$ and let λ_1' denote the first eigenvalue of the part of A in $\mathfrak{P} \ominus \mathfrak{P}_n$. We know from the classical characterization (1.3.1) that

$$\lambda_1' = \min_{\substack{p \in \mathfrak{P} \\ (p, p_i) = 0 \\ i = 1, 2, \ldots, n}} [(Ap, p)/(p, p)].$$

Moreover, since $\mathfrak{Q}_n = \mathfrak{Q} \oplus \{\mathfrak{P} \ominus \mathfrak{P}_n\}$, we have from (2.3)

$$\lambda_1' + \lambda_k^{(\infty)} \leq \lambda_k^{(n)} \qquad (k = 1, 2, \ldots). \tag{2}$$

We now note again that $(p, p_i) = (p, Pu_i) = (p, u_i)$ $(i = 1, 2, \ldots, n)$ and apply the *maximum–minimum principle* (3.2.1) to obtain

$$\lambda_{n+1}^{(0)} = \min_{\substack{u \in \mathfrak{H} \\ (u, u_i) = 0 \\ i = 1, 2, \ldots, n}} [(Au, u)/(u, u)] \leq \min_{\substack{p \in \mathfrak{P} \\ (p, u_i) = 0 \\ i = 1, 2, \ldots, n}} [(Ap, p)/(p, p)] = \lambda_1'. \tag{3}$$

Combining inequalities (2) and (3), we have

$$\lambda_{n+1}^{(0)} + \lambda_k^{(\infty)} \le \lambda_k^{(n)} \qquad (k = 1, 2, \ldots).$$

Noting that $\lambda_{n+1}^{(0)} \le 0$, we have the desired estimate (1).

Let us also note that if there are no *superpersistent eigenvectors*, the sequence $\{p_i\}$ is a *distinguished sequence*. To see this, we let $u_*^1, u_*^2, \ldots, u_*^\mu$ be the eigenvectors corresponding to a given $\lambda_*^{(0)}$. Then putting $p^i = Pu_*^i$ $(i = 1, 2, \ldots, \mu)$, we have

$$\det\{(p^i, u_*^k)\} = \det\{(p^i, p^k)\}, \tag{4}$$

which is the Gram determinant. If it were zero, there would be a nontrivial combination, $\alpha_1 p_1 + \alpha_2 p_2 + \cdots + \alpha_\mu p_\mu$, equal to zero, Since the eigenvector

$$u_* = \alpha_1 u_*^1 + \alpha_2 u_*^2 + \cdots + \alpha_\mu u_*^\mu$$

would then be orthogonal to $\mathfrak{P}$, u_* would be superpersistent, contrary to our hypothesis.

4. Stenger's Inequality

The following inequality [S7] is in a sense complementary to that of Aronszajn.

Let $\mathfrak{H}'$, $\mathfrak{H}''$, P', and P'' be as in Section 2. Here, we do not assume that A is bounded, as in the case of Aronszajn's inequality, but we retain the assumptions that A, $P'AP'$, and $P''AP''$ are in class $\mathscr{S}$. Besides compact operators, other more general operators satisfy these hypotheses, for instance if $A \in \mathscr{S}$ and either $\mathfrak{H}'$ or $\mathfrak{H}''$ is a finite-dimensional subspace of $\mathfrak{D}$.

Theorem 1. *The eigenvalues of the operators A, $P'A\,P'$, and $P''AP''$ satisfy the inequalities*

$$\lambda_1 + \lambda_{i+j} \le \lambda_i' + \lambda_j'' \qquad (i, j = 1, 2, \ldots). \tag{1}$$

Proof. We put $\mathfrak{D}' = \mathfrak{H}' \cap \mathfrak{D}$ and $\mathfrak{D}'' = \mathfrak{H}'' \cap \mathfrak{D}$. As in Theorem 2.1, for any $u \in \mathfrak{D}' \oplus \mathfrak{D}''$, $(u, u) = 1$, we can write $u = \sigma v + \tau w$, where $v \in \mathfrak{D}'$, $w \in \mathfrak{D}''$, $(v, v) = (w, w) = 1$, and $|\sigma|^2 + |\tau|^2 = 1$. If $\tau = 0$, we have $u = v$, $R(u) = R(v)$, and $\lambda_1 \le R(w)$ for any $w \in \mathfrak{D}''$. Therefore, we obtain the inequality

$$R(u) + \lambda_1 \le R(v) + R(w). \tag{2}$$

We prove now that (2) is true for $\tau \neq 0$. In fact, we have by direct computation

$$0 \leq R(u) - \lambda_1 = \sigma\bar{\sigma}[R(v) - \lambda_1] + \sigma\bar{\tau}([A - \lambda_1 I]v, w) + \bar{\sigma}\tau\overline{([A - \lambda_1 I]v, w)} + \tau\bar{\tau}[R(w) - \lambda_1].$$

Therefore, we can write

$$0 \leq [R(u) - \lambda_1]/|\tau|^2 = |\sigma/\tau|^2[R(v) - \lambda_1] + (\sigma/\tau)([A - \lambda_1 I]v, w) + \overline{(\sigma/\tau)}\overline{([A - \lambda_1 I]v, w)} + [R(w) - \lambda_1]. \qquad (3)$$

Let us now apply Lemma 1.1. We set $z = \sigma/\tau$, $\alpha = R(v) - \lambda_1$, $\beta = ([A - \lambda_1 I]v, w)$, and $\gamma = R(w) - \lambda_1$, and apply inequality (1.2) in (3) and obtain the inequality

$$[R(u) - \lambda_1]/|\tau|^2 \leq \{1 + (\sigma/\tau)(\bar{\sigma}/\bar{\tau})\}[R(v) + R(w) - 2\lambda_1].$$

Multiplying by $|\tau|^2$, we get again inequality (2). Therefore, (2) is true for every nonzero $u \in \mathfrak{D}' \oplus \mathfrak{D}''$. Now, let

$\mathfrak{U}_i' = \mathrm{sp}\{u_i', u_2', \ldots, u_i'\}$ and $\mathfrak{U}_j'' = \mathrm{sp}\{u_1'', u_2'', \ldots, u_j''\}$ $(i, j = 1, 2, \ldots)$.

Using inequality (2) and the *minimum–maximum principle*, we get

$$\lambda_{i+j} \leq \max_{u \in \mathfrak{U}_i' \oplus \mathfrak{U}_j''} R(u) \leq \max_{v \in \mathfrak{U}_i'} R(v) + \max_{w \in \mathfrak{U}_j''} R(w) - \lambda_1 = \lambda_i' + \lambda_j'' - \lambda_1,$$

which proves inequality (1).

We now give an application of our inequality (1) to the uniform estimation of the eigenvalues of the first intermediate problem as it is used in the new maximum–minimum theory (Section 7.4). Let $p \in \mathfrak{D}$ such that $(p, p) = 1$, and let $\mathfrak{H}' = \mathrm{sp}\{p\}$. Then, $\lambda_1' = R(p)$ is the only nontrivial eigenvalue of $P'AP'$ and is a Poincaré–Rayleigh–Ritz upper bound for λ_1. The numbers $\lambda_1'' \leq \lambda_2'' \ldots$ are the eigenvalues of the first intermediate problem

$$Au - (Au, p)p = \lambda u, \qquad (u, p) = 0,$$

and satisfy the inequality (see Section 7.2)

$$\lambda_j'' \leq \lambda_{j+1} \qquad (j = 1, 2, \ldots). \qquad (4)$$

Applying inequality (1) together with (4) we, have

$$0 \leq \lambda_{1+j} - \lambda_j'' \leq \lambda_1' - \lambda_1 \qquad (j = 1, 2, \ldots). \qquad (5)$$

which gives an estimate of the excess of λ_{j+1} over $\lambda_j'' = \lambda_j^{(1)}$, even if p is not an optimal choice as in Section 2.4.

5. A Procedure for Obtaining Upper Bounds

In the present section, we give an upper-bound method which requires the roots of two determinants whose matrices are both half the size of the corresponding matrix in the *Rayleigh–Ritz method* (Section 2.3). However, unlike the Rayleigh–Ritz method, here we require a rough lower bound, say μ, for the *first* eigenvalue λ_1 of the given operator A. See [S15].

We choose a set of "test functions" $v_1, v_2, \ldots, v_{2n} \in \mathfrak{D}$ in such a way that the subspace $\mathfrak{V}' = \mathrm{sp}\{v_1, v_2, \ldots, v_n\}$ is orthogonal to the subspace $\mathfrak{V}'' = \mathrm{sp}\{v_{n+1}, v_{n+2}, \ldots, v_{2n}\}$. Let V' and V'' be the orthogonal projection operators onto $\mathfrak{V}'$ and $\mathfrak{V}''$, respectively. The (nontrivial) eigenvalues

$$\Lambda_1' \le \Lambda_2' \le \cdots \le \Lambda_n'$$

and

$$\Lambda_1'' \le \Lambda_2'' \le \cdots \le \Lambda_n''$$

of

$$V'AV'u = \Lambda u$$

and

$$V''AV''u = \Lambda u$$

are given by the roots of

$$\det\{(Av_i, v_j) - \Lambda(v_i, v_j)\} = 0 \qquad (i, j = 1, 2, \ldots, n) \tag{1}$$

and

$$\det\{(Av_i, v_j) - \Lambda(v_i, v_j)\} = 0 \qquad (i, j = n+1, n+2, \ldots, 2n), \tag{2}$$

respectively (see Section 2.3).

Since $\mathfrak{V}'$ and $\mathfrak{V}''$ are both n-dimensional subspaces of $\mathfrak{D}$, we have Poincaré's inequalities (2.1.2)

$$\lambda_1 \le \Lambda_1', \quad \lambda_2 \le \Lambda_2', \ldots, \quad \lambda_n \le \Lambda_n' \tag{3}$$

and

$$\lambda_1 \le \Lambda_1'', \quad \lambda_2 \le \Lambda_2'', \ldots, \quad \lambda_n \le \Lambda_n''. \tag{4}$$

Using the notation of Section 4, we now put $\mathfrak{H}'' = \mathfrak{H} \ominus \mathfrak{V}'$ and $\mathfrak{H}' = \mathfrak{V}'$. We note first that, since $\mathfrak{H}' = \mathfrak{V}'$, we have

$$\lambda_1' = \Lambda_1', \quad \lambda_2' = \Lambda_2', \ldots, \quad \lambda_n' = \Lambda_n'. \tag{5}$$

Moreover, $\mathfrak{B}'$ is finite-dimensional, so that, by Theorem 3.3.1, the operator $P''AP''$ is in class $\mathscr{S}$. Since $\mathfrak{B}''$ is a finite-dimensional subspace of $\mathfrak{H}''$, we have again Poincaré's inequalities (2.1.2)

$$\lambda_1'' \leq \Lambda_1'', \quad \lambda_2'' \leq \Lambda_2'', \ldots, \quad \lambda_n'' \leq \Lambda_n''. \tag{6}$$

Applying inequality (4.1) to (5) and (6) and using our rough lower bound μ, we have

$$\mu + \lambda_{i+j} \leq \lambda_1 + \lambda_{i+j} \leq \lambda_i' + \lambda_j'' \leq \Lambda_i' + \Lambda_j'' \qquad (i, j = 1, 2, \ldots, n), \tag{7}$$

which we rewrite as

$$\lambda_{i+j} \leq \Lambda_i' + \Lambda_j'' - \mu \qquad (i, j = 1, 2, \ldots, n). \tag{8}$$

We see from (5), (6), and (8) that

$$\lambda_1 \leq \Lambda_1', \Lambda_1'',$$
$$\lambda_2 \leq \Lambda_2', \Lambda_2'', \Lambda_1' + \Lambda_1'' - \mu,$$
$$\lambda_3 \leq \Lambda_3', \Lambda_3'', \Lambda_1' + \Lambda_2'' - \mu, \Lambda_2' + \Lambda_1'' - \mu,$$

and so on. Therefore, we have the following set of upper bounds for the first $2n$ eigenvalues:

$$\lambda_1 \leq \min\{\Lambda_1', \Lambda_1''\}$$
$$\lambda_k \leq \min_{i+j=k} \{\Lambda_k', \Lambda_k'', \Lambda_i' + \Lambda_j'' - \mu\} \qquad (k = 2, 3, \ldots, n).$$
$$\lambda_k \leq \min_{i+j=k} \{\Lambda_i' + \Lambda_j'' - \mu\} \qquad (k = n + 1, n + 2, \ldots, 2n).$$

Note that in the *Rayleigh–Ritz method* upper bounds for $2n$ eigenvalues would require the roots of the determinant of a $2n \times 2n$ matrix, namely

$$\det\{(Av_i, v_i) - \Lambda(v_i, v_j)\} \qquad (i, j = 1, 2, \ldots, 2n),$$

whereas in (1) and (2) the matrices are $n \times n$.

6. Inequalities for Sums of Eigenvalues

We discuss here inequalities of Fan [F1] and Hersch [H3–4] concerning sums of eigenvalues, for which we give modified proofs. Moreover, we give criteria for the equality signs to hold, paralleling Weinstein's new maximum–minimum theory.

Let us begin with the following lemma.

Lemma 1. (*Invariance of the trace*) *Let $v_1, v_2, \ldots, v_n$ be orthonormal vectors in $\mathfrak{D}$ and let γ_{ij} $(i, j = 1, 2, \ldots, n)$ be scalars such that*

$$\sum_{i=1}^{n} \left(\gamma_{ij} \bar{\gamma}_{ik} \Big/ \sum_{s=1}^{n} |\gamma_{is}|^2 \right) = \delta_{jk} \qquad (j, k = 1, 2, \ldots, n). \tag{1}$$

Then, we have

$$\sum_{i=1}^{n} R\left(\sum_{j=1}^{n} \gamma_{ij} v_j \right) = \sum_{i, j, k=1}^{n} \left[\gamma_{ij} \bar{\gamma}_{ik} \Big/ \sum_{s=1}^{n} |\gamma_{is}|^2 \right] (Av_j, v_k)$$

$$= \sum_{k=1}^{n} (Av_k, v_k) = \sum_{k=1}^{n} R(v_k). \tag{2}$$

For any n-dimensional subspace $\mathfrak{V}_n \subset \mathfrak{D}$, we denote by $\mathbf{Tr}[\mathfrak{V}_n]$ the *trace* of A in $\mathfrak{V}_n$, that is,

$$\mathbf{Tr}[\mathfrak{V}_n] = \sum_{i=1}^{n} R(v_i),$$

where $\{v_1, v_2, \ldots, v_n\}$ is any orthonormal basis for $\mathfrak{V}_n$.

Theorem 1. (*Fan*) *Let $v_1, v_2, \ldots, v_n$ be any orthonormal vectors in $\mathfrak{D}$. Then, the sum of the first n eigenvalues of A satisfies*

$$\lambda_1 + \lambda_2 + \cdots + \lambda_n = \min_{v_1, v_2, \ldots, v_n} \sum_{i=1}^{n} R(v_i). \tag{3}$$

The original proof of Fan [F1] was computational rather than variational. Later, Hersch [H3–4] and Diaz and Metcalf [DM1] gave proofs based on Weyl's inequality. In order to get a more complete result (see [S5, S13]), we give a proof based on Poincaré's inequalities (2.1.2).

PROOF. We note from Lemma 1 and inequalities (2.1.2) (using the notations of Section 2.1) that

$$\lambda_1 + \lambda_2 + \cdots + \lambda_n \le \Lambda_1 + \Lambda_2 + \cdots + \Lambda_n$$

$$= \sum_{i=1}^{n} R(w_i) = \sum_{i=1}^{n} R(v_i). \tag{4}$$

By putting $v_1 = u_1, v_2 = u_2, \ldots, v_n = u_n$, we have equality in (4), thus proving (3).

Using Theorem, 2.4.4, we can add the following.

REMARK. *For any choice of* $v_1, v_2, \ldots, v_n \in \mathfrak{D}$ *such that* $\mathrm{sp}\{v_1, v_2, \ldots, v_n\} \neq \mathrm{sp}\{u_1, u_2, \ldots, u_n\}$, *we have the strict inequality*

$$\lambda_1 + \lambda_2 + \cdots + \lambda_n < \sum_{i=1}^{n} R(v_i). \tag{5}$$

Let $\mathfrak{P}_n$ denote any n-dimensional subspace of $\mathfrak{H}$ and let $\mathfrak{V}_r$ be any r-dimensional subspace of $\mathfrak{D}$ such that $\mathfrak{V}_r \perp \mathfrak{P}_n$.

Theorem 2. (*Hersch*) *For any indices* r, $n = 1, 2, \ldots,$ *we have the inequalities*

$$\lambda_{n+1} + \lambda_{n+2} + \cdots + \lambda_{n+r} \geq \min_{\mathfrak{V}_r \perp \mathfrak{P}_n} \mathbf{Tr}[\mathfrak{V}_r]. \tag{6}$$

PROOF. Using the notations of Section 7.8, we see that inequalities (7.8.7) imply

$$\lambda_{n+1} + \lambda_{n+2} + \cdots + \lambda_{n+r} \geq \lambda_1^{(n)} + \lambda_2^{(n)} + \cdots + \lambda_r^{(n)}, \tag{7}$$

where $\lambda_1^{(n)}, \lambda_2^{(n)}, \ldots, \lambda_r^{(n)}$ are the first r eigenvalues of (4.2.4). By Theorem 1, we have for $\mathfrak{V}_r \perp \mathfrak{P}_n$,

$$\lambda_1^{(n)} + \lambda_2^{(n)} + \cdots + \lambda_r^{(n)} = \min_{\mathfrak{V}_r \perp \mathfrak{P}_n} \mathbf{Tr}[\mathfrak{V}_r],$$

which combined with (7) yields inequality (6).

Let us now give our criterion for the equality sign to hold in inequality (6), or equivalently, in inequality (7).

Theorem 3. (*Stenger*) *A necessary and sufficient condition on the subspace* $\mathfrak{P}_n$ *of Theorem* 1 *for the simultaneous equalities*

$$\lambda_1^{(n)} = \lambda_{n+1}, \quad \lambda_2^{(n)} = \lambda_{n+2}, \ldots, \quad \lambda_r^{(n)} = \lambda_{n+r} \tag{8}$$

to hold is that, for every $\lambda \in (\lambda_{n+1}, \lambda_{n+r})$, *the quadratic form defined by the symmetric matrix*

$$\{(R_\lambda p_j, p_k)\} \qquad (j, k = 1, 2, \ldots, n) \tag{9}$$

have in canonical coordinates the diagonal form

$$-\mathbf{x}_1{}^2 - \mathbf{x}_2{}^2 - \cdots - \mathbf{x}_n{}^2 \tag{10}$$

and that the criterion of Theorem 7.4.1 *for* $\lambda_1^{(n)} = \lambda_{n+1}$ *hold.*

PROOF. Let us first consider the special case in which

$$\lambda_{n+1} = \lambda_{n+2} = \cdots = \lambda_{n+r}.$$

In this case, the interval $(\lambda_{n+1}, \lambda_{n+r})$ is empty, so that (10) is vacuously satisfied. By Theorem 7.4.1, we have $\lambda_1^{(n)} = \lambda_{n+1}$ if and only if the quadratic form for (9) evaluated at $\lambda_{n+1} - \varepsilon$ has the diagonal form

$$-\mathbf{x}_1{}^2 - \mathbf{x}_2{}^2 - \cdots - \mathbf{x}_{m-1}^2 + \mathbf{x}_m{}^2 + \cdots + \mathbf{x}_n{}^2. \tag{11}$$

If (9) has form (11) at $\lambda_{n+1} - \varepsilon$, then (9) has form (11) at $\lambda_{n+h} - \varepsilon$ $(h = 1, 2, \ldots, r)$ and so equalities (8) hold.

Let us now consider the case in which there exists an index k_0 such that

$$\lambda_{n+1} = \cdots = \lambda_{n+k_0} < \lambda_{n+k_0+1} \leq \cdots \leq \lambda_{n+r}. \tag{12}$$

By the same reasoning as given above, the criterion of Theorem 7.4.1 for $\lambda_1^{(n)} = \lambda_{n+1}$ yields the set of equalities

$$\lambda_2^{(n)} = \lambda_{n+2}, \ldots, \quad \lambda_{k_0}^{(n)} = \lambda_{n+k_0}.$$

Moreover, we have $\min\{j \mid \lambda_j = \lambda_{n+k_0+1}\} = n + k_0 + 1$. Now, if for every $\lambda \in (\lambda_{n+1}, \lambda_{n+r})$ the quadratic form of (9) has form (10), then by Theorem 7.8.2 each of the equalities in (8) is satisfied, so that our condition is sufficient. On the other hand, if equalities (8) hold, the set of equalities

$$\begin{aligned} \lambda_{k_0+1}^{(n)} &= \lambda_{k_0+2}^{(n-1)} = \cdots = \lambda_{n+k_0}^{(1)} = \lambda_{n+k_0+1}, \\ \lambda_{k_0+2}^{(n)} &= \lambda_{k_0+3}^{(n-1)} = \cdots = \lambda_{n+k_0+1}^{(1)} = \lambda_{n+k_0+2}, \ldots \\ \lambda_r^{(n)} &= \lambda_{r+1}^{(n-1)} = \cdots = \lambda_{n+r-1}^{(1)} = \lambda_{n+r}, \end{aligned} \tag{13}$$

must all hold simultaneously. Let us first consider

$$\lambda_{n+k_0}^{(1)} = \lambda_{n+k_0+1}, \ldots, \quad \lambda_{n+r-1}^{(1)} = \lambda_{n+r}. \tag{14}$$

Case A. If $-\infty < W_{01}(\lambda_{n+k_0+1}) < 0$, then, by Theorem 7.2.1 and the monotonicity of $W_{01}(\lambda)$, equalities (14) imply

$$-\infty < W_{01}(\lambda) < 0, \qquad \lambda \in (\lambda_{n+1}, \lambda_{n+r}). \tag{15}$$

Case B. If $W_{01}(\lambda_{n+k_0+1}) = \infty$, then, by Theorem 7.2.1, the equality $\lambda_{n+k_0}^{(1)} = \lambda_{n+k_0+1}$ means that $\lambda_{n+k_0} = \lambda_{n+k_0+1}$, which is a contradiction.

Case C. If $0 \leq W_{01}(\lambda_{n+k_0+1}) < \infty$, then, by Theorem 7.2.1, we again have $\lambda_{n+k_0} = \lambda_{n+k_0+1}$, a contradiction.

By repeating these same arguments for $W_{12}(\lambda)$, $W_{23}(\lambda)$, $\ldots$, $W_{n-1,n}(\lambda)$ as in the proof of Theorem 7.8.2, we obtain our criterion.

7. Diaz and Metcalf's Generalized Inequalities

Diaz and Metcalf used Lemma 6.1 to obtain the following nontrivial generalizations of (2.1) and (4.1), see [DM1].

Let $\mathfrak{H}^{(1)}, \mathfrak{H}^{(2)}, \ldots, \mathfrak{H}^{(n)}$ be mutually orthogonal subspaces, $\mathfrak{H} = \mathfrak{H}^{(1)} \oplus \mathfrak{H}^{(2)} \oplus \cdots \oplus \mathfrak{H}^{(n)}$, such that the projections of A to $\mathfrak{H}^{(1)}, \mathfrak{H}^{(2)}, \ldots, \mathfrak{H}^{(n)}$ are of class $\mathscr{S}$, say if A is compact. We denote by $\lambda_i^{(j)}$ and $u_i^{(j)}$ the (nontrivial) eigenvalues and eigenvectors of the corresponding projections of A ($i = 1, 2, \ldots; j = 1, 2, \ldots, n$). With these notations, we have the following.

Theorem 1. *For any indices $i_1, i_2, \ldots, i_n$, we have*

$$\lambda_1 + \lambda_2 + \cdots + \lambda_{n-1} + \lambda_{i_1+i_2+\cdots+i_n} \le \lambda_{i_1}^{(1)} + \lambda_{i_2}^{(2)} + \cdots + \lambda_{i_n}^{(n)}. \tag{1}$$

PROOF. We note that any vector in $\mathfrak{D}$ may be written as

$$u = \sum_{i=1}^{n} \alpha^{(i)} u^{(i)}.$$

where $u^{(i)} \in \mathfrak{H}^{(i)}$ ($i = 1, 2, \ldots, n$). Taking $\|u\| = 1$ for simplicity, we put $\gamma_{1j} = \alpha^{(j)}$ ($j = 1, 2, \ldots, n$) and apply Lemma 6.1 and Theorem 6.1 to obtain

$$\begin{aligned}\lambda_1 + \lambda_2 + \cdots + \lambda_{n-1} + R(u) &\le \sum_{i=2}^{n} R\left(\sum_{j=1}^{n} \gamma_{ij} u^{(j)}\right) + R\left(\sum_{j=1}^{n} \gamma_{1j} u^{(j)}\right) \\ &= \sum_{i=1}^{n} R\left(\sum_{j=1}^{n} \gamma_{ij} u^{(j)}\right) = \sum_{j=1}^{n} R(u^{(j)}),\end{aligned} \tag{2}$$

where the γ_{ij} are chosen to satisfy (6.1). Now, let $\mathfrak{U}_{i_j}^{(j)} = \operatorname{sp}\{u_1^{(j)}, u_2^{(j)}, u_{i_j}^{(j)}\}$ ($j = 1, 2, \ldots, n$) and put $\mathfrak{B}$ equal to the $(i_1 + i_2 + \cdots + i_n)$-dimensional space $\mathfrak{U}_{i_1}^{(1)} + \mathfrak{U}_{i_2}^{(2)} + \cdots + \mathfrak{U}_{i_n}^{(n)}$. Then, by (2) and Poincaré's inequality (2.2.4), we have

$$\begin{aligned}&\lambda_1 + \lambda_2 + \cdots + \lambda_{n-1} + \lambda_{i_1+i_2+\cdots+i_n} \\ &\qquad \le \lambda_1 + \lambda_2 + \cdots + \lambda_{n-1} + \max_{u \in \mathfrak{B}} R(u) \\ &\qquad \le \sum_{j=1}^{n} \max_{u^{(j)} \in \mathfrak{U}_{i_j}^{(j)}} R(u^{(j)}) = \lambda_{i_1}^{(1)} + \lambda_{i_2}^{(2)} + \cdots + \lambda_{i_n}^{(n)},\end{aligned}$$

which proves (1).

In the following, we make the additional assumption that A is bounded and put $\mu = \sup_{u \in \mathfrak{D}} R(u)$.

Theorem 2. *For any indices* $i_1, i_2, \ldots, i_n$, *we have*

$$\lambda_{i_1}^{(1)} + \lambda_{i_2}^{(2)} + \cdots + \lambda_{i_n}^{(n)} \le \lambda_{i_1+i_2+\cdots+i_n-n+1} + (n-1)\mu. \tag{3}$$

Moreover, when $\dim \mathfrak{H} = N < \infty$, *we have*

$$\lambda_{i_1}^{(1)} + \lambda_{i_2}^{(2)} + \cdots + \lambda_{i_n}^{(n)} \le \lambda_{i_1+i_2+\cdots+i_n-n+1} + \lambda_N + \lambda_{N-1} + \cdots + \lambda_{N-n+2} \tag{4}$$

where $1 \le i_1 + i_2 + \cdots + i_n - n + 1 \le N$ *and* $2 \le N - n + 2 \le N$. *These two side conditions require that* $n \le i_1 + i_2 + \cdots + i_n \le N + n - 1$ *and* $2 \le n \le N$.

PROOF. Using the same notations as above, we have

$$\sum_{j=1}^{n} R(u^{(j)}) = R\left(\sum_{j=1}^{n} \gamma_{1j} u^{(j)}\right) + \sum_{i=2}^{n} R\left(\sum_{j=1}^{n} \gamma_{ij} u^{(j)}\right)$$

$$\le R(u) + (n-1)\mu \tag{5}$$

if $\dim \mathfrak{H} = \infty$ and

$$\le R(u) + \lambda_N + \lambda_{N-1} + \cdots + \lambda_{N-n+2} \tag{6}$$

if $\dim \mathfrak{H} = N$. In the latter, we have used Theorem 6.1 with the eigenvalues enumerated in a *decreasing* order (see also Section 3.5). If we now require u to satisfy the orthogonality condition $u \perp \mathfrak{W}$, where

$$\mathfrak{W} = \operatorname{sp}\{u_1^{(1)}, u_2^{(1)}, \ldots, u_{i_1-1}^{(1)}, u_1^{(2)}, \ldots, u_{i_2-1}^{(2)}, \ldots, u_1^{(n)}, \ldots, u_{i_n-1}^{(n)}\},$$

we have, by (3.1.2), (5), and (6),

$$\lambda_{i_1}^{(1)} + \lambda_{i_2}^{(2)} + \cdots + \lambda_{i_n}^{(n)} = \sum_{j=1}^{n} \min_{u^{(j)} \perp \mathfrak{A}_{i_j-1}^{(j)}} R(u^{(j)})$$

$$= \min_{u \perp \mathfrak{W}} R(u) + (n-1)\mu$$

$$= \lambda_{i_1+i_2+\cdots+i_n-n+1} + (n-1)\mu$$

if $\dim \mathfrak{H} = \infty$ and

$$\le \lambda_{i_1+i_2+\cdots+i_n-n+1} + \lambda_N + \lambda_{N-1} + \cdots + \lambda_{N-n+2}$$

if $\dim \mathfrak{H} = N$, which proves (3) and (4).

Let us note that inequality (3) but *not* the sharper inequality (4) could also be derived directly from Aronszajn's inequality (2.1) by induction.

8. An Inequality for Sums of Operators

The inequalities due to Aronszajn (2.1) should not be confused with another set of inequalities due to Weyl [W31, p. 445], which we give in this section. Although these two results resemble each other upon a superficial inspection, they are nevertheless fundamentally different in character. As we shall see, the proof of Weyl's inequalities requires only the maximum–minimum principle, while that of Aronszajn's inequalities uses Lemma 1.1 in addition to the maximum–minimum principle. Using the results of Section 3.3, we can extend Weyl's inequalities, which he originally proved for compact operators, to operators of class $\mathscr{S}$.

Let $A = A^{[1]} + A^{[2]}$, where A, $A^{[1]}$, $A^{[2]} \in \mathscr{S}$. We denote by $\lambda_1 \leq \lambda_2 \leq \ldots$, $\lambda_1^{[1]} \leq \lambda_2^{[1]} \leq \ldots$, $\lambda_1^{[2]} \leq \lambda_2^{[2]} \leq \ldots$; $u_1, u_2, \ldots, u_1^{[1]}, u_2^{[1]}, \ldots, u_1^{[2]}, u_2^{[2]}, \ldots$; λ_∞, $\lambda_\infty^{[1]}$, $\lambda_\infty^{[2]}$, E_λ, $E_\lambda^{[1]}$, $E_\lambda^{[2]}$; $\mathfrak{D}$, $\mathfrak{D}^{[1]}$, $\mathfrak{D}^{[2]}$ the eigenvalues, orthonormal eigenvectors, infima of the essential spectra, spectral resolutions, and domains of A, $A^{[1]}$, and $A^{[2]}$, respectively. We consider again representations of the type

$$Au = \sum_{k=1} \lambda_k(u_k, u)u_k + \int_{\lambda_\infty - 0}^{\infty} \lambda \, dE_\lambda u$$

and

$$A^{[r]}u = \sum_{k=1} \lambda_k^{[r]}(u_k^{[r]}, u)u_k^{[r]} + \int_{\lambda_\infty{}^{[r]} - 0}^{\infty} \lambda \, dE_\lambda^{[r]}u \qquad (r = 1, 2).$$

Then, we have the following extension of Weyl's result.

Theorem 1. *The eigenvalues of the operators A, $A^{[1]}$ and $A^{[2]}$ satisfy*

$$\lambda_i^{[1]} + \lambda_j^{[2]} \leq \lambda_{i+j-1} \qquad (i, j = 1, 2, \ldots). \tag{1}$$

PROOF. Our proof closely follows Weyl's proof for compact operators. Let the degenerate (finite-rank) operators $A_n^{[1]}$ and $A_n^{[2]}$ be defined by

$$A_n^{[r]}u = \sum_{k=1}^{n} \lambda_k^{[r]}(u_k^{[r]}, u)u_k^{[r]} \qquad (r = 1, 2;\ n = 1, 2, \ldots). \tag{2}$$

For a fixed pair of indices (i, j), let

$$A^{\#} = A - A_{i-1}^{[1]} - A_{j-1}^{[2]}. \tag{3}$$

Denote by $R^{\#}(u)$ the Rayleigh quotient of $A^{\#}$ and by $R_n^{[r]}(u)$ the Rayleigh quotient of $A^{[r]} - A_n^{[r]}$, where $r = 1, 2$ and $n = 1, 2, \ldots$. Since

$$A^{\#} = A^{[1]} - A_{i-1}^{[1]} + A^{[2]} - A_{j-1}^{[2]}.$$

we have

$$R^{\#}(u) = R_{i-1}^{[1]}(u) + R_{j-1}^{[2]}(u).$$

From the definition of $A_n^{[r]}$ it follows that

$$\min_{u \in \mathfrak{D}} R_n^{[r]}(u) = \lambda_{n+1}^{[r]} \qquad (r = 1, 2; n = 1, 2, \ldots).$$

Therefore, we obtain

$$\begin{aligned} \lambda_i^{[1]} + \lambda_j^{[2]} &= \min_{u \in \mathfrak{D}^{[1]}} R_{i-1}^{[1]}(u) + \min_{u \in \mathfrak{D}^{[2]}} R_{j-1}^{[2]}(u) \\ &= \min_{u \in \mathfrak{D}^{[1]} \cap \mathfrak{D}^{[2]} = \mathfrak{D}} R^{\#}(u). \end{aligned} \tag{4}$$

Now, let $\mathfrak{U}_{i,j} = \mathrm{sp}\{u_1^{[1]}, u_2^{[1]}, \ldots, u_{i-1}^{[1]}, u_1^{[2]} u_2^{[2]}, \ldots, u_{j-1}^{[2]}\}$. From the definition of $A^{\#}$, we see that, for all $u \in \mathfrak{U}_{i,j}^{\perp} \cap \mathfrak{D}$, the equation $A^{\#}u = Au$ is satisfied. Therefore, we have

$$\min_{u \in \mathfrak{D}} R^{\#}(u) \le \min_{u \in \mathfrak{A}_{i,j}^{\perp} \cap \mathfrak{D}} R^{\#}(u) = \min_{u \in \mathfrak{A}_{i,j}^{\perp} \cap \mathfrak{D}} R(u). \tag{5}$$

From the maximum–minimum principle, we obtain

$$\min_{u \in \mathfrak{A}_{i,j}^{\perp} \cap \mathfrak{D}} R(u) \le \lambda_{i+j-1}. \tag{6}$$

Combining inequalities (4)–(6), we have inequality (1), which completes the proof.

The following result is easily derived from (1) by induction.

Corollary 1. *If* $A = A^{[1]} + A^{[2]} + \cdots + A^{[n]}$, *then the corresponding eigenvalues satisfy*

$$\lambda_{i_1}^{[1]} + \lambda_{i_2}^{[2]} + \cdots + \lambda_{i_n}^{[n]} \le \lambda_{i_1 + i_2 + \cdots + i_n - n + 1}. \tag{7}$$

Inequalities (1) and (7) yield elementary lower bounds to the eigenvalues of A. It may be of some interest to compare these with the lower bounds obtained by truncation in Section 6.4. Fox and Rheinboldt [FR1] have shown that, for sufficiently large truncation indices, the truncation bounds are not less than the elementary bounds. In fact, if we take for simplicity $A = A^{[1]} + A^{[2]}$ and choose truncation indices $n_1 > i$ and $n_2 > j$, we obtain (see Section 6.4 for notations)

$$\begin{aligned}
\lambda_i^{[1]} + \lambda_j^{[2]} &= \min_{\substack{u \perp \mathfrak{A}_{i-1}^{[1]} \\ u \in \mathfrak{B}_{\mathbf{n}} \cap \mathfrak{D}^{[1]}}} \frac{(A^{[1]}u, u)}{(u, u)} + \min_{\substack{u \perp \mathfrak{A}_{j-1}^{[2]} \\ u \in \mathfrak{B}_{\mathbf{n}} \cap \mathfrak{D}^{[2]}}} \frac{(A^{[2]}u, u)}{(u, u)} \\
&\leq \min_{\substack{u \perp \mathfrak{A}_{i-1}^{[1]} + \mathfrak{A}_{j-1}^{[2]} \\ u \in \mathfrak{B}_{\mathbf{n}} \cap \mathfrak{D}}} \frac{(A^{[1]}u, u)}{(u, u)} + \min_{\substack{u \perp \mathfrak{A}_{i-1}^{[1]} + \mathfrak{A}_{j-1}^{[2]} \\ u \in \mathfrak{B}_{\mathbf{n}} \cap \mathfrak{D}}} \frac{(A^{[2]}u, u)}{(u, u)} \\
&\leq \min_{\substack{u \perp \mathfrak{A}_{i-1}^{[1]} + \mathfrak{A}_{j-1}^{[2]} \\ u \in \mathfrak{B}_{\mathbf{n}} \cap \mathfrak{D}}} \frac{([A^{[1]} + A^{[2]}]u, u)}{(u, u)} \\
&= \min_{\substack{u \perp \mathfrak{A}_{i-1}^{[1]} + \mathfrak{A}_{j-1}^{[2]} \\ u \in \mathfrak{B}_{\mathbf{n}} \cap \mathfrak{D}}} \frac{(T^{\mathbf{n}}u, u)}{(u, u)},
\end{aligned}$$

which, by Weyl's inequality (3.1.2), is not greater than $\lambda_{i+j-1}^{\mathbf{n}}$.

Chapter Nine

Intermediate Problems and Perturbation Theory

1. Summary of Previous Results

While at various places (see Sections 4.1 and 5.2) we have mentioned the fact that intermediate problems are not perturbation theory, we begin the present chapter by giving a more detailed review of this question.

As rightly observed by Kato [19, p. xix], perturbation theory is not a sharply defined discipline. Usually, perturbation theory starts with a *given operator* A whose spectrum is assumed to be known either qualitatively or quantitatively. In its original form, this theory dealt with the determination of some properties of the spectrum of $A + \varepsilon B$, where ε is a small parameter and B is another given operator. Later, beginning perhaps with Weyl in 1909, important investigations were made about the change of the spectrum of A if A is replaced by $A + B$, which Riesz and Sz.-Nagy call the perturbation of the operator A [26, p. 367].

According to Friedrichs [8, p. 3], the method of perturbations deals with an operator T which is, in one sense or another, *near* an operator A whose spectral resolution is known. However, a cursory perusal of the book by Friedrichs seems to indicate that the starting point is not the operator T but A, and that T *is not given* but *is defined by* $A + B$, where B is a more or less gentle operator. As to the word *near*, whatever it means, it could not be applied to the base problem in the method of intermediate problems. For example, in the Mathieu equation (see Section 5.5), for a certain value of the parameter, a lower bound for the first eigenvalue of the given problem is 2.48 ..., while the first eigenvalue of the base problem is 0 and therefore does not belong to a very restricted neighborhood of the actual value.

Even in the case when the operator T is already in the form $A + B$, where A is a base problem (which is the starting point for the second type of intermediate problems), it was a most significant achievement of Aronszajn [A5, A8], especially in view of its various applications, to form the operators $A + B_n$ in such a way that they are intermediate between A and T (see Section 5.1). At this point, one can clearly see the difference between intermediate problems and perturbations. In fact, Kato, in his book of perturbation theory [19, p. 244], considers only the case in which $A + B$ is such that the given operator B is already degenerate. In perturbation theory, the idea of building stepping stones between a base problem and a given problem, which originated in Weinstein's first type of intermediate problems, does not occur at all. Moreover, we shall see in Section 4 that the first type of problems is even more remote from perturbation theory.

However, the following sections will again show the significant fact that the advanced developments of the theory of degenerate perturbations have been influenced by intermediate problems.

The finite-rank perturbation problem at present can be solved by both the general rule (Section 5.5), which yields numerical results, and by Aronszajn's rule (Section 5.3), which is useful in theoretical questions.

2. Two Optimum Problems for Positive Perturbations of Finite Rank

In Sections 5.2–5.5, we have determined the eigenvalues of an operator perturbed by a degenerate operator, using the methods of intermediate problems. In this section, we discuss some inequalities relating the eigenvalues of perturbed and unperturbed operators with the corresponding optimum problems, thereby paralleling Weinstein's new maximum–minimum theory. The results of this section are due to Stenger [S10].

Let D_r be a positive- (or negative-) semidefinite, degenerate operator of rank r and let $A_r = A + D_r$ [see Eq. (5.2.2)] with $\alpha_i > 0$ $(i = 1, 2, \ldots, r)$. We note again (Section 5.1) that, by a classical theorem of Weyl's [W30], the essential spectrum of A_r is the same as that of A. Moreover, we have shown in Section 5.1 that the lower part of the spectrum of A_r consists of isolated eigenvalues $\lambda_1^{(r)}, \lambda_2^{(r)}, \ldots$ satisfying the inequalities

$$\lambda_i^{(r)} \leq \lambda_{i+r} \qquad (i = 1, 2, \ldots). \tag{1}$$

We again denote by $u_1^{(r)}, u_2^{(r)}, \ldots$ the eigenvectors corresponding to $\lambda_1^{(r)}$, $\lambda_2^{(r)}, \ldots$.

Let us emphasize that the proof of (1) given in Section 5.1 shows that

(1) holds regardless of the magnitude or sign of the degenerate perturbation. This result is therefore quite different from other results in perturbation theory, which depend on the norm of the perturbation. To illustrate this fact, we consider the following simple eigenvalue problem:

$$Au + \alpha_1(u, u_1)u = \lambda u, \tag{2}$$

where A is a 3×3 matrix having eigenvalues $\lambda_1 < \lambda_2 < \lambda_3$ and corresponding eigenvectors u_1, u_2, u_3. As (2) is just the problem of the special choice (5.4.1), we see by inspection that the eigenvectors of (2) are again u_1, u_2, and u_3 with corresponding eigenvalues $\lambda_1 + \alpha_1, \lambda_2$, and λ_3. As long as $\alpha_1 < \lambda_2 - \lambda_1$, the first eigenvalue $\lambda_1^{(1)}$ of (2) is $\lambda_1 + \alpha_1$. However, if $\alpha_1 \geq \lambda_2 - \lambda_1$, then $\lambda_1^{(1)} = \lambda_2$, *regardless of how large* α_1 *is.*

Now, noting that D_r is positive, we have $(Au, u) = (A_r u, u)$ for all $u \in \mathfrak{D}$, so that the monotonicity principle (2.5.3) yields the classical complementary inequalities

$$\lambda_i \leq \lambda_i^{(r)} \qquad (i = 1, 2, \ldots,). \tag{3}$$

We are now going to investigate the question of how to choose D_r in *optimal* ways, that is, how to choose scalars α_i and vectors q_i $[(q_i, q_i) = 1]$ so that the equality sign holds in (1) or in (3).

We begin by recalling the decomposition (5.3.6). The matrix of determinant (5.2.7) is nonsymmetric, a fact which did not play any role up to now. However, in the following it is essential to have a symmetric matrix. The symmetrization here is attained by multiplying the kth column by the positive constant α_k $(k = 1, 2, \ldots, n)$. In this way, in place of (5.2.7), we obtain a new determinant

$$\det\{\alpha_k \delta_{jk} + \alpha_j \alpha_k (R_\lambda q_j, q_k)\} = \alpha_1 V_{01} \alpha_2 V_{12} \cdots \alpha_r V_{r-1,r}.$$

The new determinant differs from the original determinant by the positive factor $\alpha_1 \alpha_2 \cdots \alpha_r$. Therefore, there are no changes in the determination of the eigenvalues of A_r. Once we have Theorems 5.3.1 and 5.3.2 along with the fact (see Section 4.8) that each function $V_{j,j+1}$, $(j = 0, 1, \ldots, r-1)$ is meromorphic and monotonic on the lower part of the spectrum of A_j $(j = 0, 1, \ldots, r-1)$, we see that our problem is, *in form*, the same as the problem of the equality signs in the new maximum–minimum theory (Section 7.4) and the related inequalities (Section 7.8). Therefore, we can give immediately the following results.

Theorem 1. *For a given index i, the equality*

$$\lambda_i^{(r)} = \lambda_{i+r}$$

holds if and only if, for any admissible $\varepsilon > 0$, *the quadratic form defined by the symmetric matrix*

$$\{\alpha_k \delta_{jk} + \alpha_j \alpha_k (R_\lambda q_j, q_k)\} \qquad (j, k = 1, 2, \ldots, r) \tag{4}$$

evaluated at $\lambda = \lambda_{i+r} - \varepsilon$ *has in canonical coordinates the diagonal form*

$$-\mathbf{x}_1{}^2 - \mathbf{x}_2{}^2 - \cdots - \mathbf{x}_{m-i}^2 \pm \mathbf{x}_{m-i+1}^2 \pm \cdots \pm \mathbf{x}_{m-1}^2 + \mathbf{x}_m{}^2 + \cdots + \mathbf{x}_r{}^2,$$

where $m = \min\{j | \lambda_j = \lambda_{i+r}\}$.

Theorem 2. *For a given index i, the equality*

$$\lambda_i = \lambda_i^{(r)}$$

holds if and only if, for any admissible $\varepsilon > 0$, *the quadratic form defined by the symmetric matrix*

$$\{\alpha_k \delta_{jk} + \alpha_j \alpha_k (R_\lambda q_j, q_k)\} \qquad (j, k = 1, 2, \ldots, r)$$

evaluated at $\lambda = \lambda_i + \varepsilon$ *has in canonical coordinates the diagonal form*

$$\pm \mathbf{x}_1{}^2 \pm \mathbf{x}_2{}^2 \cdots \pm \mathbf{x}_{M-i}^2 + \mathbf{x}_{M-i+1}^2 + \cdots + \mathbf{x}_r{}^2,$$

where $M = \max\{j | \lambda_j = \lambda_i\}$.

The optimum problem solved in Theorem 1 clarifies a paper of Dolberg's [D3]. Dolberg investigates a special case of perturbations of finite rank. However, he mistakenly interprets his results as pertaining to the maximum–minimum theory. Dolberg's analysis is in two parts.

First of all, he considers a symmetric perturbation of finite rank of a positive-definite (symmetric) integral operator K having a continuous kernel. Dolberg's problem essentially can be written as

$$Ku - \sum_{j=1}^{r} [(u, Kp_j)/(Kp_j, p_j)]Kp_j = (1/\mu)u. \tag{5}$$

This equation is of type (5.2.3). To solve (5), Dolberg uses Fredholm's theory in its classical form.

Noting that K is positive and that for integral operators the eigenvalues are considered in the reciprocal sense, we denote by $\mu_1 \leq \mu_2 \leq \cdots$ and $\mu_1^{(r)} \leq \mu_2^{(r)} \leq \cdots$ the eigenvalues of the unperturbed and perturbed operators, respectively. Dolberg obtains a criterion for the equality sign in the inequality $\mu_1^{(r)} \leq \mu_{r+1}$, but only for $r + 1 = \min\{j | \mu_j = \mu_{r+1}\}$. In this special case, Dolberg's criterion resembles our criterion given in Theorem 1 if we would put $\varepsilon = 0$ (see Remark 7.2.2).

It should be noted that, even for integral operators, Dolberg's approach would not lead to a criterion for a general perturbation such as (5.2.3), since it is an essential part of his proof to have the unperturbed operator K in the degenerate perturbation.

The second part of Dolberg's paper consists of an attempt to link Eq. (5) with the maximum–minimum theory. However, this part of Dolberg's paper, which is still mentioned by Gohberg and Krein [10, English translation, p. 25] and is again repeated by Lidskii [12a, p. 71], is *essentially incorrect* since he assumes that (5) is the Euler equation of another variational problem which can be reduced in very special cases to the variational problem in the maximum–minimum theory (3.2.1), when in fact it is the Euler equation of a different problem. For a more detailed analysis, see Appendix A and [S10, SW1, W28].

3. Kuroda's Generalization of the Weinstein–Aronszajn Determinant

We now discuss a generalization of the theory of the Weinstein–Aronszajn (WA) determinant to closed operators. Kuroda's result [K10] was not included in Chapter 5 because it has no direct connection with intermediate problems, but can be considered as a significant contribution to perturbation theory which was stimulated by intermediate problems. We give here only a brief sketch, without any attempt at completeness, and refer for details to the original paper [K10] and to the book by Kato [19].

Kuroda considered a class of closed operators (which includes selfadjoint operators, see Definition A.16) and used the following theorems of operational calculus.

Let A be a closed operator with nonempty resolvent set and let λ_* be an isolated point of the spectrum A.

Definition. The algebraic multiplicity of λ_*, written $\nu(\lambda_*)$, is defined to be the dimension (possibly infinite) of the subspace

$$\mathfrak{P}_* = \{x \in \mathfrak{D}(A) \,|\, (A - \lambda_* I)^n x = 0 \quad \text{for some natural number } n\}.$$

If $\lambda_* \in \rho(A)$, we put $\nu(\lambda_*) = 0$.

Let us note that, if $\mathfrak{P}_*$ contains a nonzero vector, then λ_* is an eigenvalue. Moreover, if A is a selfadjoint operator, the algebraic and geometric multiplicities are identical (see Definition A.25). In fact, let λ_* be an isolated point of the spectrum of A. Let $\mathfrak{P}_*$ be as defined above and let $\mathfrak{E}_*$ be the (geometric) eigenspace of λ_*, namely

$$\mathfrak{E}_* = \{x \in \mathfrak{D}(A) \,|\, (A - \lambda_* I)x = 0\}.$$

Clearly, we have $\mathfrak{E}_* \subset \mathfrak{P}_*$. On the other hand, suppose $x \in \mathfrak{P}_*$ but $x \notin \mathfrak{E}_*$. Then, $x_1 = (A - \lambda_* I)x \neq 0$. Therefore, we have

$$0 \neq (x_1, x_1) = ([A - \lambda_* I]x, x_1) = (x, [A - \lambda_* I]x_1),$$

which means that

$$x_2 = (A - \lambda_* I)x_1 = (A - \lambda_* I)^2 x \neq 0.$$

Proceeding inductively, we have

$$x_n = (A - \lambda_* I)^n x \neq 0 \qquad (n = 1, 2, \ldots),$$

or $x \notin \mathfrak{P}_*$, which is a contradiction.

For the following, we assume that $\nu(\lambda_*) < \infty$. We consider the resolvent $R_\lambda = (A - \lambda I)^{-1}$. It can be shown that R_λ is meromorphic in any domain Ω containing only a finite number of isolated eigenvalues (see [19, p. 188].

Theorem 1. *Let Γ be a circle in the resolvent set for A which encloses λ_* but no other point of the spectrum. Then,*

$$P_* = (-1/2\pi i) \oint R_\lambda \, d\lambda \tag{1}$$

is the projector onto $\mathfrak{P}_*$ [19, p. 178].

For a more restricted class of bounded operators, this formula (1) goes back to Riesz and is easily verified for a symmetric, compact operator.

Since the dimension of the range of a projector is equal to the trace of the projector, we have

$$\nu(\lambda_*) = \mathbf{tr}\, P_*.$$

We now need to define "determinant" for a class of operators in Hilbert space. Let us note that if T were a positive–definite (symmetric) matrix having eigenvalues $\tau_1, \tau_2, \ldots, \tau_n$, all less than 1, we would have

$$\det(I - T) = \prod_{k=1}^{n} (1 - \tau_k) = \exp\left(\sum_{k=1}^{n} \log(1 - \tau_k) \right)$$
$$= \exp\{\mathbf{tr}[\log(I - T)]\},$$

where

$$\log(I - T) = T + \tfrac{1}{2}T^2 + \tfrac{1}{3}T^3 + \cdots.$$

Motivated by this fact, Kuroda defined a function $\mathrm{Log}(I - T)$ for any operator T having a finite trace in the following way. (Note that T has a

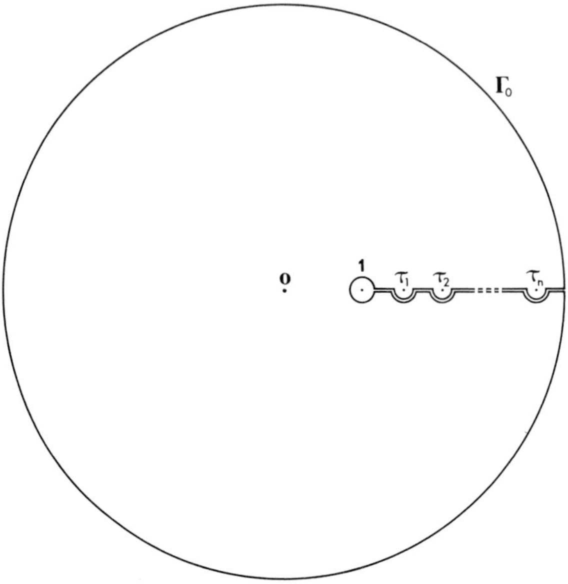

FIG. 1

finite norm and is compact but not necessarily symmetric.) For the moment, we assume that $1 \in \rho(T)$ and denote by $\{\tau_1, \tau_2, \ldots, \tau_n\}$, $\tau_1 < \tau_2 < \cdots < \tau_n$, the finite set of real eigenvalues (if any) of T that are greater than 1. Let Γ_0 be the closed contour shown in Fig. 1. The radius of the circle about zero is taken to be greater than $\|T\|$, so that the spectrum of T is entirely within the contour. In conformity with Dunford's theory of operational calculus [4], the operator $\operatorname{Log}(I - T)$ is defined by

$$\operatorname{Log}(I - T) = (-1/2\pi i) \oint_{\Gamma_0} \log(1 - \xi) R_\xi \, d\xi, \tag{2}$$

where $\log(1 - \xi)$ is taken to be real for real $\xi < 1$. This formula (2) can be easily verified for a matrix by direct computation. Since the function defined by $g(\xi) = \xi^{-1} \log(1 - \xi)$ is regular on and inside Γ_0, the operator $g(T)$ defined by

$$g(T) = (-1/2\pi i) \oint_{\Gamma_0} \xi^{-1} \log(1 - \xi) R_\xi \, d\xi$$

is bounded. From this, it follows that the operator $\operatorname{Log}(I - T) = Tg(T)$ has a finite trace, since it is the product of a bounded operator and a compact operator having finite trace.

For any T having finite trace, the *determinant of* $I - T$ is defined by

$$\det(I - T) = \begin{Bmatrix} \exp\{\mathbf{tr}[\mathrm{Log}(I - T)]\} & \text{if} \quad 1 \in \rho(T) \\ 0 & \text{if} \quad 1 \notin \rho(T) \end{Bmatrix}. \tag{3}$$

Kuroda called (3) the "infinite determinant."

Kuroda then considered any operator B having the properties: (a) $\mathfrak{D}(A) \subset \mathfrak{D}(B)$ and (b) BR_λ has finite trace for some (or equivalently for all) $\lambda \in \rho(A)$; so that B is not necessarily of finite rank but could be, for instance, any operator having finite trace. The spectrum of $A' = A + B$ is related to that of A by means of the function

$$V(\lambda) = \det(I + BR_\lambda), \tag{4}$$

which is equal to the ordinary WA determinant (5.2.6) whenever B is a symmetric operator of finite rank, say D_r [see Eq. (5.2.3)]. In fact, we have

$$\begin{aligned} \det(I + D_r R_\lambda) &= \det\{(q_i, q_j) + (D_r R_\lambda q_i, q_j)\} \\ &= \det\{\delta_{ij} + \alpha_i(R_\lambda q_i, q_j)\}, \end{aligned}$$

which is identical to (5.2.6).

The following is Kuroda's generalization of Aronszajn's rule (5.3.11).

Theorem 2. (*Kuroda*) *Let A, B, A' and $V(\lambda)$ be as above. Consider R_λ in a domain Ω in which R_λ is meromorphic and suppose that $v(\lambda)$ is finite for all $\lambda \in \Omega$. Let $\omega(\lambda_*)$ denote the order of $V(\lambda)$ at λ_* and let $v'(\lambda_*)$ denote the algebraic multiplicity of λ_* as an eigenvalue of $A' = A + B$. Then, $V(\lambda)$ is meromorphic in Ω and*

$$v'(\lambda_*) = v(\lambda_*) + \omega(\lambda_*) \qquad (\lambda_* \in \Omega). \tag{5}$$

Let us now give a sketch of Kuroda's derivation. We first note that $v'(\lambda) \equiv \infty$ in Ω if and only if $V(\lambda) \equiv 0$ in Ω. For operators of class $\mathscr{S}$, there are no nontrivial cases for which $V(\lambda) \equiv 0$, so that in the following we assume that $V(\lambda) \not\equiv 0$ in Ω.

We now derive the following relationship between the resolvents R_λ' of A' and R_λ of A for $\lambda \in \rho(A) \cap \rho(A')$, namely

$$\begin{aligned} R_\lambda' - R_\lambda &= (A + B - \lambda I)^{-1} - (A - \lambda I)^{-1} \\ &= R_\lambda'[I - (A + B - \lambda I)(A - \lambda I)^{-1}] \\ &= R_\lambda'[A - \lambda I - (A + B - \lambda I)]R_\lambda \\ &= -R_\lambda' B R_\lambda . \end{aligned} \tag{6}$$

From (6), we also have

$$R_\lambda' = R_\lambda(I + BR_\lambda)^{-1}. \tag{7}$$

Using relationships (6) and (7), Kuroda gave a rigorous deduction of the formula

$$V'(\lambda)/V(\lambda) = \mathbf{tr}\{R_\lambda - R_\lambda'\} \qquad [V(\lambda) \neq 0; \quad \lambda \in \rho(A) \cap \rho(A')]. \tag{8}$$

Let us give here a formal derivation. In view of the definitions of $V(\lambda)$ and of "determinant" (2), (3), we write

$$\begin{aligned} V'(\lambda)/V(\lambda) &= d\,\{\mathbf{tr}[\mathrm{Log}(I + BR_\lambda)]\}/d\lambda \\ &= \mathbf{tr}\{d\,[\mathrm{Log}(I + BR_\lambda)]/d\lambda\} \\ &= \mathbf{tr}\{(I + BR_\lambda)^{-1}BR_\lambda^2\} \\ &= \mathbf{tr}\{R_\lambda(I + BR_\lambda)^{-1}BR_\lambda\} = \mathbf{tr}\{R_\lambda' BR_\lambda\} \\ &= \mathbf{tr}\{R_\lambda - R_\lambda'\}. \end{aligned}$$

Now, letting Γ and λ_* be as in Theorem 1, we have

$$\begin{aligned} \omega(\lambda_*) &= (1/2\pi i) \oint_\Gamma [V'(\lambda)/V(\lambda)]\, d\lambda \\ &= (1/2\pi i) \oint_\Gamma \mathbf{tr}\{R_\lambda - R_\lambda'\}\, d\lambda \\ &= \mathbf{tr}(1/2\pi i) \oint_\Gamma [R_\lambda - R_\lambda']\, d\lambda \\ &= -v(\lambda_*) + v'(\lambda_*) \end{aligned}$$

which yields (5).

In [10, English translation, p. 156] of 1969 and [10, p. 198] of 1965 Krein claims Theorem 2 of 1961 for the case of bounded operators. However, neither Theorem 2 nor Aronszajn's rule of 1948–1950 can be found in any of Krein's previous papers. Krein's failure to obtain the rule may have been due to the fact that he and his collaborators assume that the *order* of the classical Fredholm determinant gives automatically the *multiplicity* of the eigenvalue. Lidskii [12a, p. 198] in 1970 still accepts this assumption and applies it to a perturbation problem in the case of an integral operator having a nonsymmetric kernel, thereby reaching unsubstantiated conclusions. (See Example 2, Appendix A.)

Let us again emphasize that in intermediate problems Aronszajn's rule cannot be applied (numerically) in many cases, and, in those cases in which it is applicable, it can be replaced by one or more of the other rules which are discussed in Chapters 4–6.

4. Connections between Two Types of Intermediate Problems and the Irreducible Case

Kuroda [K10] was the first to point out certain remarkable connections between the first and second types of intermediate problems. Later, Fichera [6] independently considered similar questions. In view of the importance of these questions, we shall give an analysis below which will show that such connections exist only in a limited sense, depending on whether the operator is bounded or unbounded, and therefore it cannot be said that the first type of problem can be reduced to the second type (see [S8, S11]). Moreover, it will be shown that, in those cases in which such a reduction is possible, it leads to serious limitations in applications.

In order to reduce intermediate problems of the first type, (4.2.4), to those of the second type, Kuroda considered

$$Q_n A Q_n u = \lambda u, \tag{1}$$

where again $Q_n = I - P_n$, P_n a projector onto the n-dimensional space $\mathfrak{P}_n$ spanned by $p_1, p_2, \ldots, p_n$. Kuroda then wrote (1) as

$$(I - P_n)A(I - P_n)u = \lambda u,$$

which can be expanded to

$$(A - Q_n AP_n - P_n A)u = \lambda u. \tag{2}$$

If we have $\mathfrak{P}_n \subset \mathfrak{D}$, which is always true *if A is bounded*, then

$$-Q_n AP_n - P_n A \tag{3}$$

may be considered to be a perturbation of finite rank. However, let us emphasize that the perturbation given by (3) is not necessarily of rank n but rather of rank r, where $n \leq r \leq 2n$. Therefore, this procedure leads to a WA determinant (5.2.6) of an $r \times r$ matrix. On the other hand, the original problem (1) when treated directly and *not* as a perturbation leads to a Weinstein determinant (4.3.4) of an $n \times n$ matrix. In fact, the range of (3) is spanned by $p_1, p_2, \ldots, p_n, Q_n Ap_1, Q_n Ap_2, \ldots, Q_n Ap_n$, or equivalently, since $Q_n = I - P_n$, by

$$p_1, p_2, \ldots, p_n, Ap_1, Ap_2, \ldots, Ap_n. \tag{4}$$

Of course, it could occasionally happen, say by taking $p_i = u_i (i = 1, 2, \ldots, n)$, that we have in (4) only n independent vectors, but such cases are obviously exceptional.

Another interesting fact which Kuroda proved is that, for all values of $n = 1, 2, \ldots$, the identity

$$V_{0r}(\lambda) = (-\lambda)^n W_{0n}(\lambda) \tag{5}$$

holds. Here r, as above, is the rank of the operator (3). For the proof of (5), see [10] or [19], where the indices r and n are not put in evidence.

To illustrate (5), we compare the two determinants for $n = 1$. In this case, we have

$$V(\lambda) = \det \begin{bmatrix} 1 - (R_\lambda Ap, p) + (Ap, p)(R_\lambda p, p) & -(R_\lambda p, p) \\ (R_\lambda Ap, Ap) - (Ap, p)(R_\lambda p, Ap) & -1 + (R_\lambda p, Ap) \end{bmatrix}. \tag{6}$$

Let us note that already here the matrix for $V(\lambda)$ is nonsymmetric, while for all $n \geq 1$ the matrix of the Weinstein determinant is *always* symmetric. This simplest case shows that the replacement of $W(\lambda)$ by $V(\lambda)$, even if possible, cannot be considered a simplification for any purposes and $W(\lambda)$ continues to play a significant role. The computation of $V(\lambda)$ requires the computation of three terms in addition to $(R_\lambda p, p)$, which alone already yields all results and makes the computation of $V(\lambda)$ superfluous. For this reason, (5) or (6) may be called the *hydra formula* because the reduction of $W(\lambda)$ to $V(\lambda)$ makes $W(\lambda)$ reappear accompanied by three more heads. It would indeed be difficult to guess from (6) that $V(\lambda) = -\lambda W(\lambda)$. Nevertheless, it can be verified directly by somewhat elaborate computations as follows.

$$\begin{aligned}
V(\lambda) &= \det \begin{bmatrix} 1 - (R_\lambda Ap, p) + (Ap, p)(R_\lambda p, p) & -(R_\lambda p, p) \\ (R_\lambda Ap, Ap) - (Ap, p)(R_\lambda p, Ap) & -1 + (R_\lambda p, Ap) \end{bmatrix} \\
&= -1 + (R_\lambda Ap, p) - (Ap, p)(R_\lambda p, p) + (R_\lambda p, Ap) \\
&\quad - (R_\lambda Ap, p)(R_\lambda p, Ap) + (Ap, p)(R_\lambda p, p)(R_\lambda p, Ap) \\
&\quad + (R_\lambda p, p)(R_\lambda Ap, Ap) - (R_\lambda p, p)(Ap, p)(R_\lambda p, Ap) \\
&= (p, p) - (Ap, p)W(\lambda) - (R_\lambda Ap, p)(R_\lambda p, Ap) \\
&\quad + W(\lambda)(R_\lambda Ap, Ap) \\
&= (p, p) - (R_\lambda Ap, p)(R_\lambda p, Ap) + W(\lambda)[(R_\lambda Ap, Ap) - (Ap, p)] \\
&= (R_\lambda Ap, p) - \lambda(R_\lambda p, p) - (R_\lambda Ap, p)(R_\lambda p, Ap) \\
&\quad + W(\lambda)[R_\lambda[Ap - \lambda p], Ap) + \lambda(R_\lambda p, Ap) - (Ap, p)] \\
&= (R_\lambda Ap, p)[1 - (R_\lambda p, Ap)] - \lambda W(\lambda) \\
&\quad + W(\lambda)[(p, Ap) + \lambda(R_\lambda p, Ap) - (Ap, p)]
\end{aligned}$$

$$= (R_\lambda Ap, p)[(p, p) - (R_\lambda p, Ap)] - \lambda W(\lambda)$$
$$+ \lambda W(\lambda)[(R_\lambda p, Ap)]$$
$$= (R_\lambda Ap, p)[(R_\lambda Ap, p) - \lambda(R_\lambda p, p) - (R_\lambda p, Ap)]$$
$$- \lambda W(\lambda) + \lambda W(\lambda)(R_\lambda p, Ap)$$
$$= - \lambda W(\lambda)(R_\lambda Ap, p) - \lambda W(\lambda) + \lambda W(\lambda)(R_\lambda p, Ap)$$
$$= - \lambda W(\lambda).$$

From this simple case, it is clear that reduction of the first to the second type of problem only complicates the computations. It is no wonder that no numerical applications of this reduction have been given or, to our knowledge, even attempted.

Besides the question of numerical results, it is important to note that the perturbation approach seemingly does not lend itself to theoretical applications such as the new maximum–minimum theory (see Chapter 7), as can be seen in the following.

THE IRREDUCIBLE CASE. If we study in more detail the case of an unbounded operator of class $\mathscr{S}$, the question of reduction becomes even more complicated. In this case, Eq. (2) cannot even be formed unless we limit the vectors p_i in the domain of A. Otherwise, if some p_i were not in $\mathfrak{D}$, then the domain of (3) would not be dense and the theory could not be applied.

The restriction of the vectors $p_1, p_2, \ldots, p_n$ to $\mathfrak{D}$ renders decomposition (2) inapplicable to an extension of the maximum–minimum principle to unbounded selfadjoint operators. In fact, the maximum–minimum principle requires that the p_i be arbitrary elements in $\mathfrak{H}$, having nothing whatsoever to do with the domain of the operator, since the only purpose of the p_i is to define orthogonality conditions [see (7.1.1)].

On the other hand, by not trying to stretch $Q_n A Q_n$ on the Procrustean bed of perturbation theory even if $\mathfrak{P}_n \not\subset \mathfrak{D}$ and/or $\mathfrak{Q}_n \not\subset \mathfrak{D}$, we have shown that $Q_n A Q_n$ is a selfadjoint operator having a dense domain $\mathfrak{P}_n \oplus (\mathfrak{D} \cap \mathfrak{Q}_n)$ (see Theorem A.2).

5. A Transformation for Positive Operators

Fichera [6] (see also the later papers [L5, L6]) considered a *positive, compact operator A* and established the following connection between the eigenvalue problem of the first type

$$(I - P_n)Au = \lambda u, \qquad P_n u = 0, \tag{1}$$

and the problem

$$Av - A^{1/2}P_n A^{1/2}v = \xi v. \tag{2}$$

The latter problem is obviously motivated by the special case (6.6.4). The assumption of positiveness is of course necessary since we use $A^{1/2}$. Compactness could be replaced by boundedness here since the purpose of this assumption is to guarantee that $A^{1/2}P_n A^{1/2}v$ is well defined for an arbitrary choice of $\mathfrak{P}_n$. However, we retain Fichera's original assumption, since compactness is necessary for the proof of Theorem 2.

Theorem 1. *Every eigenvalue for the problem* (1) *is an eigenvalue for* (2). *Conversely, every nonzero eigenvalue for* (2) *is an eigenvalue for* (1).

PROOF. Let λ_* be an eigenvalue for (1) with eigenvector u_*. If we put $v_* = A^{1/2}u_*$, Eq. (1) becomes

$$A^{1/2}v_* - P_n A^{1/2}v_* = \lambda_* u_*. \tag{3}$$

Operating on both sides of (3) with $A^{1/2}$, we obtain

$$Av_* - A^{1/2}P_n A^{1/2}v_* = \lambda_* v_*$$

so that λ_* is an eigenvalue for (2) with eigenvector $v_* = A^{1/2}u_*$. On the other hand, let $\xi_* \neq 0$ be an eigenvalue for (2) with eigenvector v_*. Then, we have

$$v_* = (1/\xi_*)A^{1/2}(A^{1/2} - P_n A^{1/2})v_*. \tag{4}$$

Let $u_* = (1/\xi_*)(A^{1/2} - P_n A^{1/2})v_*$. Then, from (4) we have $v_* = A^{1/2}u_*$ or $A^{1/2}v_* = Au_*$, so that (4) becomes

$$A^{1/2}[Au_* - P_n Au_* - \xi_* u_*] = 0.$$

Since $A^{1/2}$ is strictly positive, u_* must be an eigenvector for (1) with eigenvalue ξ_*, which completes the proof.

Since the operator $-A^{1/2}P_n A^{1/2}$ is of finite rank, we can apply the results of Chapter 5. However, we must keep in mind the fact that the eigenfunctions of (1) and (2) are different.

Stenger [S14] established the following connection between the Weinstein determinants for (1) and (2), which is superficially similar to but essentially different from (4.5).

Theorem 2. *The determinants* (4.3.4) *and* (5.2.6) *corresponding to* (1) *and* (2) *satisfy*

$$V_{0n}(\xi) = (-\xi)^n W_{0n}(\xi) \tag{5}$$

for all ξ.

PROOF. Assuming for simplicity

$$(p_i, p_j) = \delta_{ij} \qquad (i, j = 1, 2, \ldots, n),$$

we write (2) as

$$Au - \sum_{i=1}^{n} (A^{1/2} p_i, u) A^{1/2} p_i = \xi u.$$

Here, the WA determinant is given by

$$V_{0n}(\xi) = \det\{\delta_{ij} - (R_\xi A^{1/2} p_i, A^{1/2} p_j)\} \qquad (i, j = 1, 2, \ldots, n).$$

Note that here the matrix is symmetric since $\alpha_i = -1$ $(i = 1, 2, \ldots, n)$. Since A is a positive, compact operator, for any real ξ, $\xi \neq \lambda_j$ $(j = 1, 2, \ldots)$, and for any $u \in \mathfrak{H}$ we have

$$\begin{aligned}
R_\xi A^{1/2} u &= R_\xi \sum_{j=1}^{\infty} \lambda_j^{1/2} (u, u_j) u_j \\
&= \sum_{j=1}^{\infty} \lambda_j^{1/2} (u, u_j) R_\xi u_j \\
&= \sum_{j=1}^{\infty} \lambda_j^{1/2} (u, u_j)(\lambda_j - \xi)^{-1} u_j \\
&= \sum_{j=1}^{\infty} \lambda_j^{1/2} (u, [\lambda_j - \xi]^{-1} u_j) u_j \\
&= \sum_{j=1}^{\infty} \lambda_j^{1/2} (u, R_\xi u_j) u_j \\
&= \sum_{j=1}^{\infty} \lambda_j^{1/2} (R_\xi u, u_j) u_j \\
&= A^{1/2} R_\xi u.
\end{aligned}$$

Therefore, we can write

$$\begin{aligned}
\delta_{ij} - (R_\xi A^{1/2} p_i, A^{1/2} p_j) &= (p_i, p_j) - (R_\xi A p_i, p_j) \\
&= (R_\xi [A - \xi I] p_i, p_j) - (R_\xi A p_i, p_j) \\
&= -\xi (R_\xi p_i, p_j) \qquad (i, j = 1, 2, \ldots, n),
\end{aligned}$$

so that we obtain

$$V_{0n}(\xi) = (-\xi)^n W_{0n}(\xi) \tag{6}$$

for ξ not an eigenvalue of A. However, since $V(\xi)$ and $W(\xi)$ are both meromorphic functions of ξ, Eq. (6) holds for all ξ.

6. Other Theoretical Results

In closing, let us mention several important theoretical ramifications of intermediate problems which have not been discussed in the present monograph, namely Aronszajn's and Smith's work on *functional completion* [AS1], Colautti's theorem on the *hypercompleteness of Weinstein's sequences of harmonic functions* [C1], Payne's proof of *Weinstein's conjecture* [P1], Aronszajn's study of *analytic functions with positive real part*, and Weinberger's *extension of the classical Sturm–Liouville theory* [W4].

Appendix A

In this appendix, we shall catalog some of the basic definitions, notations, and standard results in functional analysis which are used throughout the book. The proofs of most of the results mentioned here as well as detailed discussions of the general theory may be found in texts such as [4, 19, 26]. We do, however, give proofs of a few less readily available theorems. No completeness of the exposition on our part is intended and our purpose is merely to provide a convenient reference for the reader.

Definition 1. $\mathfrak{H}$ is a real (or complex) *Hilbert space* if the following properties hold:

1. $\mathfrak{H}$ is a vector space over the reals (or complexes);
2. There is a real (or complex) function (u, v) defined for every pair $u, v \in \mathfrak{H}$ such that (a) $(u, u) \geq 0$ for all $u \in \mathfrak{H}$, with equality if and only if $u = 0$; (b) $(u + v, w) = (u, w) + (v, w)$ for $u, v, w \in \mathfrak{H}$; (c) $(\alpha u, v) = \alpha(u, v)$ for $u, v \in \mathfrak{H}$ and α real (or complex); (d) $(u, v) = \overline{(v, u)}$. (The bar denotes complex conjugation where applicable.)
3. $\mathfrak{H}$ is complete with respect to the norm defined by $\|u\| = (u, u)^{1/2}$ [4, p. 1773].

We shall use the symbol $x_j \to x_0$ to denote convergence with respect to the above norm (strong convergence) of a sequence $\{x_j\}$ in $\mathfrak{H}$ to a vector $x_0 \in \mathfrak{H}$. The topological properties of subsets of $\mathfrak{H}$ are with respect to this convergence, unless otherwise specified.

Definition 2. Let $\mathfrak{S}$ be a subset of $\mathfrak{H}$. The *orthogonal complement* of $\mathfrak{S}$, written $\mathfrak{S}^\perp$, consists of all $u \in \mathfrak{H}$ such that $(u, v) = 0$ for every $v \in \mathfrak{S}$.

REMARK. The set $\mathfrak{S}^{\perp}$ is a closed subspace of $\mathfrak{H}$.

Definition 3. The *codimension of* $\mathfrak{S}$, written codim $\mathfrak{S}$, is the dimension of $\mathfrak{S}^{\perp}$.

Definition 4. A is said to be a *linear operator* from a subspace $\mathfrak{D}$ into $\mathfrak{H}$ if, for every $u \in \mathfrak{D}$, we have $Au \in \mathfrak{H}$ and for all $u, v \in \mathfrak{D}$ and all scalars α, β, we have $A(\alpha u + \beta v) = \alpha Au + \beta Av$.

Definition 5. The *domain* of A, written $\mathfrak{D}(A)$, or $\mathfrak{D}$ for short, consists of all $u \in \mathfrak{H}$ for which Au is defined.

Definition 6. The *range* of A, written $\mathfrak{R}(A)$, consists of all $v \in \mathfrak{H}$ for which there exists $u \in \mathfrak{D}$ such that $v = Au$. The *rank* of A is dim $\mathfrak{R}(A)$.

Definition 7. The *null space* of A, written $\mathfrak{N}(A)$, consists of all $u \in \mathfrak{D}(A)$ such that $Au = 0$. The *nullity* of A is given by $v(A) = \dim \mathfrak{N}(A)$.

Definition 8. A subspace $\mathfrak{D}$ of $\mathfrak{H}$ is said to be *dense* in $\mathfrak{H}$ if the closure $\overline{\mathfrak{D}}$ of $\mathfrak{D}$ equals $\mathfrak{H}$.

Definition 9. A is said to be *symmetric* if $(Au, v) = (u, Av)$ for all $u, v \in \mathfrak{D}(A)$.

From now on, A denotes a linear operator defined on a *dense* subspace $\mathfrak{D}$ of $\mathfrak{H}$.

Definition 10. Let $\mathfrak{D}^* = \{v \in \mathfrak{H} \mid (Au, v)$ is continuous in u for all u in $\mathfrak{D}\}$. For $v \in \mathfrak{D}^*$, let w be the element uniquely determined by the equation $(Au, v) = (u, w)$ for all u in $\mathfrak{D}$. Then, the *adjoint* of A, written A^*, is defined on $\mathfrak{D}^*$ by the equation $A^*v = w$. [4, p. 1188.]

Definition 11. An operator A is said to be *selfadjoint* if $\mathfrak{D} = \mathfrak{D}^*$ and $Au = A^*u$ for all $u \in \mathfrak{D}$. (Most of the book deals with selfadjoint operators.)

REMARK. Selfadjoint implies symmetric.

Definition 12. A is said to be *bounded above* (or *below*) if there exists a real number γ such that $(Au, u) \leq \gamma(u, u)$ [or $\gamma(u, u) \leq (Au, u)$] for all $u \in \mathfrak{D}(A)$. An operator which is bounded either above or below is said to be *semibounded*, and one which is bounded both above and below is said to be *bounded*.

Definition 13. Let $\mathfrak{S}$ be the set of all $u \in \mathfrak{H}$ such that $(u, u) = 1$. A linear operator K is said to be *compact* if $K\mathfrak{S}$ is a compact subset of $\mathfrak{H}$.

Definition 14. A sequence x_j in $\mathfrak{H}$ is said to *converge weakly* to x_0 (written $x_j \xrightarrow{w} x_0$) if, for every $y \in \mathfrak{H}$, $(x_j, y) \to (x_0, y)$.

REMARK. If K is a compact operator and $x_j \xrightarrow{w} x_0$, then $Kx_j \to Kx_0$.

Definition 15. A is said to be *positive-definite* (or just *positive*) if $0 < (Au, u)$ for all $u \in \mathfrak{D}(A)$ $(u \neq 0)$, *positive-semidefinite* (or *semipositive* or *nonnegative*) if $0 \leq (Au, u)$ for all $u \in \mathfrak{D}(A)$ $(u \neq 0)$, *negative-definite* if $(Au, u) < 0$ for all $u \in \mathfrak{D}(A)$ $(u \neq 0)$, and *negative-semidefinite* if $(Au, u) \leq 0$ for all $u \in \mathfrak{D}(A)$ $(u \neq 0)$.

Theorem 1. *If A is any one of the types of operators given in Definition* 15, *then* (a) *A is symmetric if $\mathfrak{H}$ is complex, or* (b) *A can be replaced by an equivalent symmetric operator if $\mathfrak{H}$ is real and $\mathfrak{D} \subset \mathfrak{D}^*$.*

PROOF. Suppose $\mathfrak{H}$ is complex and A is positive-definite. Let $x, y \in \mathfrak{D}(A)$ and let α be any scalar. Then, we have, for $x + \alpha y \neq 0$.

$$\begin{aligned} 0 &< (A[x + \alpha y], x + \alpha y) \\ &= (Ax, x) + (Ax, \alpha y) + \alpha(Ay, x) + (\alpha Ay, \alpha y) \\ &= (Ax, x) + \bar{\alpha}(Ax, y) + \alpha\overline{(x, Ay)} + |\alpha|^2(Ay, y) \\ &= (Ax, x) + |\alpha|^2(Ay, y) + \bar{\alpha}(Ax, y) + \alpha\overline{(x, Ay)}. \end{aligned} \tag{1}$$

Letting $\alpha = i$, then, from (1), we must have $-i(Ax, y) + i\overline{(x, Ay)}$ equal to real number. Therefore,

$$\begin{aligned} 0 &= \mathbf{Re}\{(Ax, y) - \overline{(x, Ay)}\} \\ &= \mathbf{Re}(Ax, y) - \mathbf{Re}(x, Ay). \end{aligned}$$

Letting $\alpha = 1$, again from (1) we must have

$$\begin{aligned} 0 &= \mathbf{Im}\{(Ax, y) + \overline{(x, Ay)}\} \\ &= \mathbf{Im}(Ax, y) - \mathbf{Im}(x, Ay). \end{aligned}$$

Therefore, $(Ax, y) = (x, Ay)$, for all $x, y \in \mathfrak{D}(A)$, which means that A is symmetric. The proofs for positive-semidefinite, etc., are similar. Now, let us suppose that $\mathfrak{H}$ is real and again that A is positive. Since A is densely defined, it has an adjoint A^* with domain $\mathfrak{D}^*$. We can write A as

$$\begin{aligned} A &= \tfrac{1}{2}(A + A^*) + \tfrac{1}{2}(A - A^*) \\ &= B + C. \end{aligned}$$

For every $u \in \mathfrak{D}$, we have

$$\begin{aligned}(Cu, u) &= \tfrac{1}{2}(Au, u) - \tfrac{1}{2}(A^*u, u) \\ &= \tfrac{1}{2}(Au, u) - \tfrac{1}{2}(u, Au) \\ &= \tfrac{1}{2}(u, A^*u) - \tfrac{1}{2}(u, Au) = -(u, Cu) \\ &= -(Cu, u),\end{aligned}$$

so that $(Cu, u) = 0$. Therefore, the quadratic form for A is equal to that for B. However, B is symmetric, since

$$\begin{aligned}(Bu, v) &= \tfrac{1}{2}(Au, v) + \tfrac{1}{2}(A^*u, v) \\ &= \tfrac{1}{2}(u, A^*v) + \tfrac{1}{2}(u, Av) \\ &= (u, Bv) \qquad (u, v \in \mathfrak{D}).\end{aligned}$$

which proves (b).

REMARK. In view of Theorem 1, whenever we consider a positive, etc., operator we mean positive symmetric operator implicitly.

Definition 16. A is said to be *closed* if, given any sequence $\{x_n\}$ in $\mathfrak{D}(A)$ such that $x_n \to x$ and $Ax_n \to y$, it follows that $x \in \mathfrak{D}(A)$ and $Ax = y$.

Closed-Graph Theorem. *A closed linear operator mapping a Hilbert space into a Hilbert space is bounded* [19, p. 166; 4, p. 57; 26, p. 306].

REMARK. If A is selfadjoint, then A is closed.

Lemma 1. *Let $\mathfrak{M}$ be a closed subspace having finite codimension in $\mathfrak{H}$. If $\mathfrak{V}$ is dense in $\mathfrak{H}$, then $\mathfrak{V} \cap \mathfrak{M}$ is dense in $\mathfrak{M}$* [GK1; 11, p. 103].

Definition 17. A linear operator P is said to be an *orthogonal projection operator* (or *projector*) from $\mathfrak{H}$ onto a closed subspace $\mathfrak{P}$ if, for every $u \in \mathfrak{H}$, we have $Pu \in \mathfrak{P}$ and $u - Pu \in \mathfrak{P}^\perp$ [26, p. 266].

REMARK. An orthogonal projection operator P is symmetric and $P^2 = P$.

Definition 18. A is said to be *degenerate* or of *finite rank* if dim $\mathfrak{R}(A) < \infty$.

Definition 19. Let A be a linear operator and let P be an orthogonal projector onto a closed subspace $\mathfrak{P}$ of $\mathfrak{H}$. The *part of A in $\mathfrak{P}$* is defined to be the restriction of PA to vectors $u \in \mathfrak{P} \cap \mathfrak{D}(A)$. The *projection of A* (or compression of A) is defined to be the operator PAP, where $\mathfrak{D}(PAP) = \{u \in \mathfrak{H} \mid Pu \in \mathfrak{D}(A)\}$.

Theorem 2. *Let T be any selfadjoint operator on a subspace $\mathfrak{D}$ dense in $\mathfrak{H}$. Let $\mathfrak{X}$ be any finite-dimensional subspace of $\mathfrak{H}$ and let $\mathfrak{Y}$ be the orthogonal complement of $\mathfrak{X}$, $\mathfrak{Y} = \mathfrak{X}^{\perp}$. Let Y denote the orthogonal projection operator onto $\mathfrak{Y}$. Then the operator $T_0 = YTY$ is selfadjoint.*

PROOF. Let $\mathfrak{D}_0$ denote the domain of T_0. Since $Yx = 0$ for all $x \in \mathfrak{X}$ and since $\mathfrak{H} = \mathfrak{X} \oplus \mathfrak{Y}$, it follows that $\mathfrak{D}_0 = \mathfrak{X} \oplus (\mathfrak{Y} \cap \mathfrak{D})$. The fact that $\mathfrak{D}_0$ is dense in $\mathfrak{H}$ follows from Lemma 1. Since T_0 is symmetric, it suffices to show that $\mathfrak{D}_0 = \mathfrak{D}_0{}^*$, where $\mathfrak{D}_0{}^*$ denotes the domain of the adjoint of T_0 (see Definition 10). First of all, $u \in \mathfrak{D}_0$ implies $Yu \in \mathfrak{D}$ $(= \mathfrak{D}^*$, since T is selfadjoint). Therefore, (Tv, Yu) is continuous for all $Yv \in \mathfrak{D}$, so that $(YTYv, u)$ is continuous for all $v \in \mathfrak{D}_0$. This means that $u \in \mathfrak{D}_0{}^*$, and therefore $\mathfrak{D}_0 \subset \mathfrak{D}_0{}^*$. On the other hand, let $u \in \mathfrak{D}_0{}^*$. Then, $(YTYv, u) = (TYv, Yu)$ is continuous for all $v \in \mathfrak{D}_0$, or equivalently, (Tw, Yu) is continuous for all $w \in \mathfrak{D} \cap \mathfrak{Y}$. It remains to show that (Tw, Yu) is continuous for all $w \in \mathfrak{D}$, which would imply that $Yu \in \mathfrak{D}^* = \mathfrak{D}$, and, therefore, that $u \in \mathfrak{D}$. Let $\mathfrak{X}_1 = \{x \in \mathfrak{X} \mid \exists y \in \mathfrak{Y} \text{ for which } x + y \in \mathfrak{D}\}$. Clearly, $\mathfrak{X}_1$ is a subspace (not necessarily proper) of $\mathfrak{X}$, and, therefore, finite-dimensional, say s-dimensional. It can be shown that $\mathfrak{X}_1 = \mathfrak{X}$ (see the end of the proof), but this fact is not used here. Let $x_1, x_2, \ldots, x_s$ be any orthonormal basis for $\mathfrak{X}_1$ and let $y_1, y_2, \ldots, y_s$ be any elements in $\mathfrak{Y}$ such that $z_j = x_j + y_j \in \mathfrak{D}$ $(j = 1, 2, \ldots, s)$. Let $\mathfrak{Z} = \text{sp}\{z_1, z_2, \ldots, z_s\}$. Then, we can write $\mathfrak{D} = \mathfrak{Z} + (\mathfrak{D} \cap \mathfrak{Y})$.

Now, let

$$\mathfrak{W}_0 = \mathfrak{D} \cap \mathfrak{Y}$$

and

$$\mathfrak{W}_j = \text{sp}\{z_1, z_2, \ldots, z_j\} \oplus \mathfrak{W}_0 \qquad (j = 1, 2, \ldots, s).$$

Observe that $\mathfrak{W}_s = \mathfrak{D}$ is obtained from $\mathfrak{W}_0$ by adding a finite-dimensional space. Our procedure is to add one dimension at a time. We have already shown that (Tw, Yu) is continuous on $\mathfrak{W}_0$. Suppose that (Tw, Yu) is continuous on $\mathfrak{W}_k$ $(0 \leq k < s)$. Let w_{k+1} denote any element in $\mathfrak{W}_{k+1}$. Then, we can write $w_{k+1} = \alpha z_{k+1} + w_k$, where $w_k \in \mathfrak{W}_k$. By construction, w_k is orthogonal to x_{k+1} and therefore

$$\begin{aligned} |(z_{k+1}, w_k)| &= |(x_{k+1} + y_{k+1}, w_k)| = |(y_{k+1}, w_k)| \\ &\leq \|y_{k+1}\| \, \|w_k\| = [\|z_{k+1}\|^2 - 1]^{1/2} \|w_k\| \\ &= [1 - (1/\|z_{k+1}\|^2)]^{1/2} \|z_{k+1}\| \, \|w_k\| \\ &\leq (1 - \varepsilon_{k+1}) \|z_{k+1}\| \, \|w_k\| \end{aligned}$$

for some ε_{k+1} $(0 < \varepsilon_{k+1} < 1)$, where ε_{k+1} depends only on x_{k+1} and y_{k+1}. Now we have, by direct computation,

$$\begin{aligned}\|w_{k+1}\|^2 &\geq |\alpha|^2\|z_{k+1}\|^2 - 2|\alpha|\,|(w_k, z_{k+1})| + \|w_k\|^2\\ &\geq |\alpha|^2\|z_{k+1}\|^2 - 2|\alpha|(1-\varepsilon_{k+1})\|z_{k+1}\|\,\|w_k\| + \|w_k\|^2 \qquad (2)\\ &= [|\alpha|\,\|z_{k+1}\| - \|w_k\|]^2 + 2\varepsilon_{k+1}|\alpha|\,\|w_k\|\,\|z_{k+1}\|.\end{aligned}$$

This inequality (2) means that

$$2|\alpha|\,\|w_k\|\,\|z_{k+1}\| \leq \|w_{k+1}\|^2/\varepsilon_{k+1} \qquad (3)$$

and

$$|\alpha|^2\|z_{k+1}\|^2 - 2|\alpha|\,\|w_k\|\,\|z_{k+1}\| + \|w_k\|^2 \leq \|w_{k+1}\|^2. \qquad (4)$$

Combining the inequalities (3) and (4), we see that

$$|\alpha|^2\|z_{k+1}\|^2 + \|w_k\|^2 \leq [1 + (1/\varepsilon_{k+1})]\|w_{k+1}\|^2. \qquad (5)$$

Now letting $\eta_{k+1} = [1 + (1/\varepsilon_{k+1})]^{1/2}$, we obtain from (5) the inequalities

$$|\alpha|\,\|z_{k+1}\| \leq \eta_{k+1}\|w_{k+1}\|, \qquad \|w_k\| \leq \eta_{k+1}\|w_{k+1}\|. \qquad (6)$$

Therefore, from the finite-dimensionality of $\mathfrak{Z}$ and the continuity of (Tw, Yu) on $\mathfrak{W}_k$, there exist constants γ and β_k such that

$$\begin{aligned}|(Tw_{k+1}, Yu)| &\leq |\alpha|\,|(Tz_{k+1}, Yu)| + |(Tw_k, Yu)|\\ &\leq |\alpha|\gamma\|z_{k+1}\| + \beta_k\|w_k\|. \qquad (7)\end{aligned}$$

Using the inequalities (6) in (7), we obtain

$$|(Tw_{k+1}, Yu)| \leq \beta_{k+1}\|w_{k+1}\|,$$

where $\beta_{k+1} = \eta_{k+1}(\gamma + \beta_k)$, and therefore, (Tw, Yu) is continuous on $\mathfrak{W}_{k+1}$ $(k = 0, 1, \ldots, s-1)$. Since $\mathfrak{D} = \mathfrak{Z} + (\mathfrak{D} \cap Y) = \mathfrak{W}_s$, this means that $Yu \in \mathfrak{D} = \mathfrak{D}^*$. Since $\mathfrak{D}_0 = \mathfrak{X} \oplus (\mathfrak{D} \cap \mathfrak{Y})$, it follows that $u \in \mathfrak{D}_0$, and therefore we have $\mathfrak{D}_0 = \mathfrak{D}_0^*$, which proves the theorem.

As mentioned above, we can show that $\mathfrak{X}_1 = \mathfrak{X}$. Suppose that $\mathfrak{X}_1 \neq \mathfrak{X}$. Then, there is some $x_0 \in \mathfrak{X}$ $(x_0 \neq 0)$ for which $(x_0, x_i) = 0$ $(i = 1, 2, \ldots, s)$. By construction, therefore, x_0 is orthogonal to every element in $\mathfrak{D}$. However, this is impossible, since $\mathfrak{D}$ is dense in $\mathfrak{H}$, so we have $\mathfrak{X} = \mathfrak{X}_1$.

Since the above theorem was given in [S9], a number of generalizations have appeared in the literature (see [G5, H5, H6, V1]).

Theorem 3. *The part of T in $\mathfrak{Y}$, say T_1, is selfadjoint.*

PROOF. The domain of T_1, which we write as $\mathfrak{D}_1$, is just $\mathfrak{D} \cap \mathfrak{Y}$, which, as above, is dense in $\mathfrak{Y}$. In view of the above proof, we only have to show that the domain of T_1^* is contained in $\mathfrak{D}_1$, that is, $\mathfrak{D}_1^* \subset \mathfrak{D}_1$. To this end, let $u \in \mathfrak{D}_1^*$. Then, $(T_1 v, u) = (YTv, u)$ is continuous for all $v \in \mathfrak{D}_1 = \mathfrak{D} \cap \mathfrak{Y}$. Since the part of T in $\mathfrak{Y}$ is an operator from a subspace of $\mathfrak{Y}$ into $\mathfrak{Y}$, we have $u = Yu$ and $v = Yv$. Therefore, we can conclude that (Tv, u) is continuous for all $v \in \mathfrak{D} \cap \mathfrak{Y}$. However, we have shown in the above proof that, moreover, (Tv, u) is continuous for all $v \in \mathfrak{D}$. Since T is selfadjoint, this means that $u \in \mathfrak{D} \cap \mathfrak{Y} = \mathfrak{D}_1$, which concludes the proof.

Definition 20. A is said to be *one-to-one* if $Au = Av$ implies $u = v$.

Definition 21. Let A be one-to-one. The *inverse* of A, written A^{-1}, is the operator defined on $\mathfrak{R}(A)$ by $A^{-1}(Au) = u$.

Let I denote the *identity operator* on $\mathfrak{H}$, $Iu = u$ for all $u \in \mathfrak{H}$, and let O denote the *zero operator* on $\mathfrak{H}$, $Ou = 0$ for all $u \in \mathfrak{H}$.

Definition 22. Let λ be a real or complex number. Suppose $A - \lambda I$ is one-to-one. Then, the inverse of $A - \lambda I$ is called the *resolvent* of A, written $R_\lambda(A)$ or usually, more simply, R_λ.

Definition 23. The *resolvent set* of A, written $\rho(A)$, is the set of all (real or complex) λ such that $A - \lambda I$ has dense range and a bounded inverse.

REMARK. If A is selfadjoint, then $\lambda \in \rho(A)$ if and only if $\mathfrak{R}(A - \lambda I) = \mathfrak{H}$ [11, p. 75; 26, p. 415].

Definition 24. The *spectrum* of A, written $\sigma(A)$, is the set of all λ not in $\rho(A)$.

REMARK. If A is closed, then $\rho(A)$ is an open set and $\sigma(A)$ is a closed set in the complex plane [4, p. 1187].

Definition 25. A number λ is said to be an *eigenvalue* of A if there exists a nonzero $u \in \mathfrak{D}(A)$ such that $Au = \lambda u$. Such a vector u is called an *eigenvector* corresponding to λ. The subspace generated by all eigenvectors corresponding to λ is called the *eigenspace* corresponding to λ. The dimension $\mu(\lambda)$ of the eigenspace is called the (*geometric*) *multiplicity* of λ.

The spectrum of A may be decomposed in the following way.

Definition 26. The *discrete spectrum* $\sigma_d(A)$ is the set of all isolated eigenvalues of A each of which has finite multiplicity [19, p. 187].

Definition 27. The *essential spectrum* $\sigma_e(A)$ is the set of all λ such that $\lambda \in \sigma(A)$ but $\lambda \notin \sigma_d(A)$ [19, 243].

EXAMPLE 1. Let $K(s, t)$ belong to $\mathscr{L}^2([0, 1] \times [0, 1])$, let $K(s, t) = K(t, s)$, a.e., and let the operator K be defined on $\mathscr{L}^2[0, 1]$ by

$$(Ku)(s) = \int_0^1 K(s, t)u(t)\,dt.$$

The spectrum of the operator K consists of finitely or infinitely many eigenvalues $\lambda_1^- \leq \lambda_2^- \leq \cdots < 0$, $\lambda_1^+ \geq \lambda_2^+ \geq \cdots > 0$ and the point $\lambda = 0$, which may or may not be an eigenvalue. For instance, if $K(s, t)$ is degenerate, we have

$$K(s, t) = \sum_{i=1}^{r} \alpha_i(s)\beta_i(t).$$

Then, $\lambda = 0$ is an eigenvalue of infinite multiplicity. On the other hand, if the operator K is positive-definite, then $\lambda = 0$ is not an eigenvalue. In any case, the essential spectrum will consist of the point $\lambda = 0$ and the discrete spectrum will consist of $\lambda_1^-, \lambda_2^-, \ldots, \lambda_1^+, \lambda_2^+, \ldots$.

EXAMPLE 2. Let K be the integral operator defined by

$$(Ku)(s) = \int_{-1}^{1} (1 + 3st + t)u(t)\,dt.$$

The eigenvalue problem

$$\lambda Ku = u$$

has a single *simple* eigenvalue, namely, $\lambda = \frac{1}{2}$. Nevertheless, $\lambda = \frac{1}{2}$ is a *double* zero of the classical Fredholm determinant

$$d(\lambda) = (1 - 2\lambda)^2.$$

This example, due to Smithies [27a], shows that the order of the zero of Fredholm's determinant *does not always give the multiplicity* of the corresponding eigenvalue. See the end of Section 9.3 for additional remarks.

EXAMPLE 3. Let H denote the multiplication operator defined for the functions $u(s)$, $u \in \mathscr{L}^2[0, 1]$ by

$$Hu = su(s).$$

For any complex λ, the resolvent $R_\lambda(H)$ is defined on the range of $H - \lambda I$ by

$$R_\lambda(H)u = [1/(s - \lambda)]u(s).$$

For $\lambda \notin [0, 1]$, $R_\lambda(H)$ is a bounded operator defined on all of $\mathscr{L}^2[0, 1]$. For $\lambda \in [0, 1]$, $R_\lambda(H)$ is an unbounded operator on the dense subspace of $\mathscr{L}^2[0, 1]$ consisting of all functions of the form $(s - \lambda)f(s)$, where $f \in \mathscr{L}^2[0, 1]$. Therefore, we have $\sigma(H) = [0, 1]$. Since there are no isolated points in $\sigma(H)$, the discrete spectrum of H is empty and the essential spectrum of H is exactly $\sigma(H)$.

REMARK. The spectrum of an operator is sometimes decomposed into point spectrum, continuous spectrum, and residual spectrum (see [19, p. 514]), but these concepts are not used in the present book.

Theorem 4. *Let $\mathfrak{H}_1$ be a closed subspace of $\mathfrak{H}$ and let $\mathfrak{H}_2 = \mathfrak{H}_1{}^{\perp}$. Let A_1 and A_2 be selfadjoint operators which are defined on $\mathfrak{D}_1$ and $\mathfrak{D}_2$ dense in $\mathfrak{H}_1$ and $\mathfrak{H}_2$ and which have ranges in $\mathfrak{H}_1$ and $\mathfrak{H}_2$, respectively. Let P_1 and P_2 denote the orthogonal projection operators onto $\mathfrak{H}_1$ and $\mathfrak{H}_2$ and let A be defined on $\mathfrak{D}_1 + \mathfrak{D}_2$ by $Au = A_1 P_1 u + A_2 P_2 u$. Then, A is selfadjoint and $\sigma(A) = \sigma(A_1) \cup \sigma(A_2)$.* [FR1, p. 430].

PROOF. We see immediately by definition that $A^* = A_1{}^*P_1 + A_2{}^*P_2 = A_1P_1 + A_2P_2 = A$. For any λ, we have

$$\mathfrak{R}(A - \lambda I) = \mathfrak{R}(A_1 - \lambda P_1) \oplus \mathfrak{R}(A_2 - \lambda P_2). \tag{8}$$

We know that, since A is selfadjoint, $\lambda \in \rho(A)$ if and only if $\mathfrak{R}(A - \lambda I) = \mathfrak{H}$. It follows by (8) that $\lambda \in \rho(A)$ if and only if $\mathfrak{R}(A_1 - \lambda P_1) = \mathfrak{H}_1$ and $\mathfrak{R}(A_2 - \lambda P_2) = \mathfrak{H}_2$. Therefore, $\lambda \in \rho(A)$ if and only if $\lambda \in \rho(A_1) \cap \rho(A_2)$, which proves the theorem.

Theorem 5. *If A is a selfadjoint operator, then all elements in $\sigma(A)$ are real* [4, p. 1193].

Hereafter, let A denote a selfadjoint operator.

Theorem 6. *Let $\lambda_1, \lambda_2, \ldots \in \sigma_d(A)$. Then, there exists $u_1, u_2, \ldots \in \mathfrak{D}(A)$, $\|u_i\| = 1$, such that $Au_i = \lambda_i u_i$ $(i = 1, 2, \ldots)$. If $\lambda_i \neq \lambda_j$, then $(u_i, u_j) = 0$.*

Definition 28. The set $\{E_\lambda\}$ is said to be a *spectral family* on the line $(-\infty, \infty)$ if the following properties hold: (a) for every real λ, E_λ is an orthogonal projection operator; (b) $E_\lambda E_\mu = E_\lambda$ for $\lambda \leq \mu$; (c) $E_{\lambda+0} = E_\lambda$; (d) $E_\lambda \to 0$ for $\lambda \to -\infty$, and $E_\lambda \to I$ for $\lambda \to \infty$.

From (b), it follows that $(E_\lambda p, p)$ is a nondecreasing function of λ. (In our notations, $\lambda + 0$ denotes the right-hand limit and $\lambda - 0$ denotes the left-hand limit.)

Theorem 7. *Every selfadjoint operator A has the representation*

$$A = \int_{-\infty}^{\infty} \lambda \, dE_\lambda,$$

which means

$$(Au, v) = \int_{-\infty}^{\infty} \lambda \, d(E_\lambda u, v) \qquad (u \in \mathfrak{D}, \quad v \in \mathfrak{H}),$$

where $\{E_\lambda\}$ *is a spectral family uniquely determined by the operator* A*; each* E_λ *is permutable with* A *as well as with all bounded operators that permute with* A [26, p. 320].

EXAMPLE 4. Every selfadjoint compact operator K has the representation

$$Ku = \sum_{j=1}^{\infty} \lambda_j^{-}(u, u_j^{-})u_j^{-} + \sum_{j=1}^{\infty} \lambda_j^{+}(u, u_j^{+})u_j^{+},$$

where

$$\lambda_1^{-} \leq \lambda_2^{-} \leq \cdots \leq 0 \leq \cdots \leq \lambda_2^{+} \leq \lambda_1^{+}.$$

Therefore, we have

$$(Ku, v) = \sum_{j=1}^{\infty} \lambda_j^{-}(u, u_j^{-})(u_j^{-}, v) + \sum_{j=1}^{\infty} \lambda_j^{+}(u, u_j^{+})(u_j^{+}, v).$$

Definition 29. The spectral family determined by A is called the *resolution of the identity* of A.

Theorem 8. *If* $\lambda \in \sigma_d(A)$*, then the operator* $E(\lambda)$ *given by* $E(\lambda) = E_\lambda - E_{\lambda-0}$ *is the orthogonal projection operator onto the subspace* $\mathfrak{E}_\lambda$ *consisting of all solutions* u *of the equation* $Au = \lambda u$.

Theorem 9. E_λ *remains constant on any interval contained in* $\rho(A)$ [26, p. 361].

Theorem 10. *Let* T *be a closed operator densely defined in* $\mathfrak{H}$ *and let* K *be a compact operator on* $\mathfrak{H}$. *Then* $\sigma_e(T) = \sigma_e(T + K)$ [19, p. 244].

Let Ω be the square $-\pi/2 < x, y < \pi/2$ and let $g(x, y; \xi, \eta)$ be the Green's function for

$$\begin{aligned} \Delta u &= f \quad \text{in } \Omega \\ u &= 0 \quad \text{on } \partial\Omega. \end{aligned}$$

Let G be the operator defined on $\mathscr{L}_2(\Omega)$ by

$$Gf(x, y) = -\iint_{\Omega} g(x, y; \xi, \eta) f(\xi, \eta)\, d\xi\, d\eta.$$

Theorem 11. *If* $f \in \mathscr{L}_2(\Omega)$, Gf *is continuous in* $\overline{\Omega}$. *If* f *is bounded,* Gf *possesses continuous first derivatives in* Ω. *If the first derivatives of* f *are bounded in* Ω, Gf *possesses continuous second derivatives in* Ω.

Theorem 12. G *is a negative-definite, symmetric operator.*

REMARK. The singular part of $g(x, y; \xi, \eta)$ is the logarithmic kernel $s(x, y; \xi, \eta) = -(1/2\pi) \log [(x - \xi)^2 + (y - \eta)^2]^{1/2}$ (see also Section 6.6).

Some Addenda

I. *General Solutions for Buckling and Vibration Equations.* In his work on intermediate problems Weinstein showed that the general solution for the buckling equation, $\Delta\,\Delta w + \lambda\,\Delta w = 0$, can be written as

$$w = v + p$$

where v and p satisfy

$$\Delta v + \lambda v = 0$$
$$\Delta p = 0.$$

Similarly for the vibration equation, $\Delta\,\Delta w - \lambda w = 0$, the general solution can be written as

$$w = v + \tilde{v}$$

where v and $\tilde{v}$ satisfy

$$\Delta v + \sqrt{\lambda} v = 0$$
$$\Delta \tilde{v} - \sqrt{\lambda} \tilde{v} = 0.$$

A corresponding decomposition also occurs for the equation of a vibrating clamped plate under tension (see Section 3 of Appendix B). Previous to these results only the decomposition of the solution of the polyharmonic equation, $\Delta^n w = 0$, was known.

II. *Weinstein's Inequality.* Corollary 4.4.2 was first given by Weinstein [W9] as the following inequality between the eigenvalues of plates and membranes.

Let $\omega_1, \omega_2, \ldots$ be the eigenvalues of the vibrating membrane problem

$$\Delta u + \omega u = 0 \quad \text{in} \quad \Omega$$
$$u = 0 \quad \text{on} \quad \partial\Omega$$

and let $\lambda_1, \lambda_2, \ldots$ be the eigenvalues of the vibrating clamped plate problem

$$\Delta\,\Delta w - \lambda w = 0 \quad \text{in} \quad \Omega$$
$$w = \partial w/\partial n = 0 \quad \text{on} \quad \partial\Omega.$$

Then we have the strict inequalities

$$\omega_n^2 < \lambda_n \qquad (n = 1, 2, \ldots).$$

This and other important inequalities for eigenvalues of membranes and plates were later proved by Diaz, Payne, and Weinberger without using intermediate problems.

III. *Dolberg's Variational Problem* (see Section 9.2). Dolberg [D3] considers the variational problem

$$\min\,(Ku, u) \qquad [(Ku, Ku) = 1;\ (u, Kp_i) = 0;\ (i = 1, 2, \ldots, n)] \tag{1}$$

where K is a *positive* compact integral operator. This problem can be reduced under severe restrictions to the classical maximum–minimum principle (3.2.1). However, Dolberg gives

$$Ku - \sum_{i=1}^{n} [(u, Kp_i)/(Kp_i, p_i)]Kp_i = (1/\mu)u \tag{2}$$

as the corresponding eigenvalue problem. This is of course absurd since the orthogonality conditions $(u, Kp_i) = 0$ mean that (2) is just $Ku = (1/\mu)u$ (see [S10, SW1, W28]). Gohberg–Krein [10, p. 45; p. 25 of English translation] and Lidskii [12a, p. 71] continue to claim that Dolberg gave the new maximum minimum theory.

IV. According to Miranda's recent book (*Partial Differential Equations of Elliptic Type*, 2nd ed., p. 122, Springer-Verlag, New York, 1970) results similar to Weyl's Second Lemma have been obtained earlier by Cacciopolli, Cimino, Zaremba, Nikodym and Sobolev.

V. *Nonsymmetric Perturbations.* In a forthcoming paper by A. Weinstein a procedure will be given for the determination of the eigenvalues of a compact symmetric operator which is perturbed by a nonsymmetric operator of finite rank. The eigenvalues and their (geometric) multiplicities will be obtained by the method of general choice (Section 4.7). Let us note that Kuroda's extension of Aronszajn's rule yields only the algebraic multiplicity. The new results will be applied to Fredholm's determinants and to other subjects.

Appendix B

In this appendix, we give tables of upper and lower bounds for eigenvalues which have been obtained by the methods of this monograph. It is not intended that these results should be exhaustive, but rather that they provide a sufficiently representative sample to demonstrate the wide applicability of the methods. We have selected for the most part those problems which seem to be of interest for applications in engineering, quantum physics, and applied mathematics.

For the sake of brevity, we shall simply state the problem, give the appropriate references, and indicate the sections in which the corresponding methods are described. Unless otherwise specified, the upper bounds were obtained by the Rayleigh–Ritz method (Section 2.3).

The lower bounds for some eigenvalues in the buckling and vibrations (also under tension) of clamped plates have been obtained by using Weinstein's distinguished choice. The higher eigenvalues in Table I have been computed by Aronszajn's rule. In the other tables the computations have been made by means of other techniques (special or general choice, truncation, etc.) discussed in the text.

1. Buckling of a Clamped Plate

The method of intermediate problems was introduced by Weinstein on the following problem (see Section 4.12).

$$\begin{aligned} \Delta\,\Delta u + \mu\,\Delta u &= 0 \qquad \text{in} \quad \Omega, \\ u = \partial u/\partial n &= 0 \qquad \text{on} \quad \partial\Omega, \\ \Omega: \qquad -\pi/2 < x,\ y &< \pi/2. \end{aligned}$$

He obtained the inequalities

$$5.30362 \le \mu_1 \le 5.31173.$$

2. Vibration of a Clamped Plate (Tables I–III†)

Differential equation and boundary conditions:

$$\Delta\,\Delta u = \lambda u \qquad \text{in } \Omega,$$
$$u = \partial u/\partial n = 0 \qquad \text{on } \partial\Omega,$$
$$\Omega: \quad -a/2 < x < a/2, \quad -b/2 < y < b/2.$$

TABLE I

SQUARE PLATE[a]

Eigenvalue number	Base problem	Circum-scribed circle	Square plate: Lower bound	Square plate: Upper bound	Inscribed circle
1	4	4.2849	13.2820	13.3842	17.13965
2	25	18.5627	55.240	56.561	74.25078
3	25	18.5627	55.240	56.561	74.25078
4	64	49.9478	120.007	124.074	199.7913
5	100	49.9478	177.67	182.14	199.7913
6	100	64.9511	178.3	184.5	259.8044
7	169	106.9193	277.42	301.55	427.6772
8	169	106.9193	277.42	301.55	427.6772
9	289	151.9438	454	477	607.7753
10	289	151.9438	454	477	607.7753
11	324	293.7572	488	548	1175.029
12	400	293.7572	600.840	621.852	1175.029
13	400	326.0292	601.569	646.939	1304.117

[a] Side length π. See Section 4.12. From [W24].

† In Tables II and III, the superscripts for the lower bounds are the truncation indices (n_1, n_2) of Section 6.4 and the superscript for the upper bounds is the dimension of the space used in the Rayleigh–Ritz method (Section 2.3).

TABLE II

SQUARE PLATE[a]

Eigenvalue number	Functions even with respect to x and even with respect to y: Lower bound $\lambda^{20,28}$	Upper bound Λ^{50}	Functions even with respect to x, even with respect to y, and even with respect to interchange of x and y: Lower bound $\lambda^{17,33}$	Upper bound Λ^{50}
1	13.287	13.294	13.292	13.294
2	177.52	177.75	179.36	179.43
3	179.23	179.44	495.68	497.05
4	492.71	497.14	980.74	981.26
5	977.65	979.64	1585.6	1593.0
6	980.15	981.30	3203.1	3244.6
7	1551.4	1584.4	3281.7	3283.5
8	1572.7	1593.5	4292.4	4316.8
9	3046.5	3246.8	6677.7	6810.2
10	3277.1	3281.8	8328.4	8333.3
11	3278.6	3283.7	9850.5	9918.3
12	4215.8	4307.9	10629.0	11590.0
13	4216.5	4318.4	13048.0	13509.0
14	5309.5	6796.8	15661.0	17764.0
15	5312.7	6816.2	17749.0	19932.0

[a] Side length π. See Section 6.4. From [BFS2].

TABLE III

RECTANGULAR PLATE—FUNCTIONS EVEN WITH RESPECT TO x AND EVEN WITH RESPECT TO y[a]

ν	$b/a = 1.00$		$b/a = 1.25$		$b/a = 1.50$		$b/a = 2.00$		$b/a = 4.00$		$b/a = 8.00$	
	$\lambda_\nu^{20,28}$	λ_ν^{50}	$\lambda_\nu^{20,28}$	λ_ν^{50}	$\lambda_\nu^{20,28}$	λ_ν^{50}	$\lambda_\nu^{20,28}$	λ_ν^{50}	$\lambda_\nu^{20,28}$	λ_ν^{50}	$\lambda_\nu^{20,28}$	λ_ν^{50}
1	80.89_4	80.93_4	55.80_7	55.83_2	45.56_1	45.58_0	37.74_4	37.75_4	32.48_3	32.48_6	31.54_9	31.55_0
2	$1080._8$	$1082._1$	497.1_8	497.9_2	276.0_7	276.5_8	125.0_2	125.2_8	44.06_1	44.12_1	33.76_4	33.78_2
3	$1091._1$	$1092._4$	$1015._2$	$1016._1$	980.4_8	981.2_0	474.0_4	475.8_6	77.50_1	77.86_7	38.76_4	38.86_3
4	$2999._7$	$3026._6$	$2037._3$	$2054._2$	$1296._0$	$1299._7$	948.9_4	949.3_9	152.2_7	153.8_5	47.64_0	48.00_9
5	$5952._0$	$5964._1$	$2546._5$	$2552._0$	$1612._5$	$1624._9$	$1259._9$	$1266._9$	297.0_4	301.4_9	61.84_0	63.02_5
6	$5967._2$	$5974._2$	$5141._8$	$5246._1$	$3369._3$	$3442._1$	$1409._6$	$1416._8$	549.7_5	559.4_1	83.29_9	86.29_3
7	$9444._9$	$9645._9$	$5808._8$	$5814._9$	$4139._1$	$4152._2$	$2026._3$	$2066._9$	921.6_9	921.8_3	115.6_9	120.7_7
8	$9574._8$	$9701._2$	$7869._9$	$7971._9$	$5731._1$	$5735._6$	$3405._5$	$3422._2$	959.9_9	975.8_4	162.5_8	170.2_1
9	$1854_7.$	$1976_7.$	$8351._9$	$8370._5$	$7077._8$	$7149._6$	$3493._2$	$3666._0$	987.6_5	989.5_3	228.1_0	238.4_2
10	$1995_1.$	$1998_0.$	$1231_9.$	$1272_5.$	$7164._2$	$7455._5$	$5658._0$	$5660._5$	$1125._1$	$1134._5$	316.9_2	330.2_3
11	$1996_0.$	$1999_1.$	$1259_7.$	$1336_8.$	$9962._9$	$1033_4.$	$5872._5$	$6406._4$	$1331._1$	$1376._2$	434.2_8	450.9_1
12	$2566_6.$	$2622_7.$	$1825_8.$	$1971_1.$	$1030_7.$	$1049_4.$	$6368._8$	$6537._0$	$1572._6$	$1607._4$	585.8_5	606.3_1
13	$2567_1.$	$2629_1.$	$1969_4.$	$2106_5.$	$1270_5.$	$1515_7.$	$7084._4$	$7113._4$	$1584._9$	$1743._3$	776.8_4	802.8_4
14	$3232_4.$	$4137_9.$	$2102_8.$	$2349_7.$	$1401_7.$	$1677_3.$	$7770._5$	$8059._2$	$1797._0$	$2273._8$	915.5_2	915.5_8
15	$3234_4.$	$4149_8.$	$2314_6.$	$2411_3.$	$1873_2.$	$1956_8.$	$8647._7$	$1093_2.$	$2172._3$	$2520._4$	930.2_8	931.5_3

[a] $a = 2.00$. See Section 6.4. From [BFS2].

3. Vibration of a Clamped Plate Under Tension (Table IV)

Differential equation and boundary conditions:

$$\Delta\,\Delta u - \tau\,\Delta u = \lambda u \quad \text{in} \quad \Omega,$$
$$u = \partial u/\partial n = 0 \quad \text{on} \quad \partial\Omega,$$
$$\Omega: \quad -\pi/2 < x,\ y < \pi/2.$$

TABLE IV

SQUARE PLATE UNDER TENSION[a]

Tension τ	Lower bounds $\lambda_1^{(1)}$	$\lambda_1^{(2)}$	$\lambda_1^{(3)}$	Upper bound Λ
5	24.982	25.222	25.236	25.509
10	36.639	36.845	36.862	37.443
15	48.084	48.253	48.284	49.261
20	59.289	59.452	59.491	61.008
30	81.651	81.760	81.809	84.372
50	125.43	125.56	125.59	130.85
100	225.56	225.63	225.65	246.58
200	443.15	443.24	443.25	477.58

[a] Side length π. From [WC1]. For $\tau = 0$, see Tables I–III.

In view of the closeness of the bounds $\lambda_1^{(1)}$, $\lambda_1^{(2)}$, $\lambda_1^{(3)}$ given by the first, second, and third intermediate problems, one would conjecture that the lower bounds are closer to the actual values than the Rayleigh–Ritz upper bound Λ in this case.

4. Rectangular Cantilever Plates (Table V)

Differential equation and boundary conditions:

$$\Delta\,\Delta u = \lambda u \qquad \text{in} \quad \Omega,$$

$$u = \partial u/\partial n = 0 \qquad \text{on} \quad \Gamma_1 \quad \text{(clamped side);}$$

$$\sigma\,\Delta u + (1-\sigma)\frac{\partial^2 u}{\partial n^2} = 0$$

$$\frac{\partial(\Delta u)}{\partial n} + (1-\sigma)\frac{d}{ds}\frac{\partial^2 u}{\partial n\,\partial t} = 0 \qquad \text{on} \quad \Gamma_2 \quad \text{(free sides);}$$

$$\begin{aligned} \Omega&: \quad -a/2 < x < a/2, \quad -b/2 < y < b/2,\\ \Gamma_1&: \quad y = -b/2,\\ \Gamma_2&: \quad x = \pm a/2, \quad y = b/2. \end{aligned}$$

Here, σ is Poisson's ratio, $\partial/\partial n$ is the exterior normal derivative, $\partial/\partial t$ is the tangential derivative, and $\partial/\partial s$ is the arc length derivative.

TABLE V

RECTANGULAR PLATE—FUNCTIONS EVEN WITH RESPECT TO x[a]

	$b/a = 0.125$		$b/a = 0.250$		$b/a = 0.500$		$b/a = 1.000$	
ν	$\lambda_\nu^{50,46}$	λ_ν^{50}	$\lambda_\nu^{50,46}$	λ_ν^{50}	$\lambda_\nu^{50,46}$	λ_ν^{50}	$\lambda_\nu^{50,46}$	λ_ν^{50}
1	$3.122_8 \times 10^{+3}$	$3.160_0 \times 10^{+3}$	$1.941_6 \times 10^{+2}$	$1.970_5 \times 10^{+2}$	$1.197_7 \times 10^{+1}$	$1.225_1 \times 10^{+1}$	$7.355_3 \times 10^{-1}$	$7.579_2 \times 10^{-1}$
2	$3.979_0 \times 10^{+3}$	$4.188_3 \times 10^{+3}$	$4.420_0 \times 10^{+2}$	$4.870_2 \times 10^{+2}$	$9.526_7 \times 10^{+1}$	$1.042_5 \times 10^{+2}$	$2.723_4 \times 10^{+1}$	$2.853_3 \times 10^{+1}$
3	$7.300_5 \times 10^{+3}$	$8.055_5 \times 10^{+3}$	$1.792_0 \times 10^{+3}$	$2.047_9 \times 10^{+3}$	$4.634_9 \times 10^{+2}$	$4.792_4 \times 10^{+2}$	$4.389_4 \times 10^{+1}$	$4.650_6 \times 10^{+1}$
4	$1.495_8 \times 10^{+4}$	$1.729_9 \times 10^{+4}$	$6.467_7 \times 10^{+3}$	$7.371_9 \times 10^{+3}$	$8.956_3 \times 10^{+2}$	$9.936_5 \times 10^{+2}$	$1.657_8 \times 10^{+2}$	$1.842_9 \times 10^{+2}$
5	$3.125_7 \times 10^{+4}$	$3.683_1 \times 10^{+4}$	$7.673_6 \times 10^{+3}$	$7.962_8 \times 10^{+3}$	$1.082_8 \times 10^{+3}$	$1.166_9 \times 10^{+3}$	$2.268_7 \times 10^{+2}$	$2.360_1 \times 10^{+2}$
6	$6.211_0 \times 10^{+4}$	$7.405_1 \times 10^{+4}$	$9.246_8 \times 10^{+3}$	$9.886_0 \times 10^{+3}$	$3.031_7 \times 10^{+3}$	$3.386_6 \times 10^{+3}$	$5.306_5 \times 10^{+2}$	$5.919_6 \times 10^{+2}$
7	$1.110_4 \times 10^{+5}$	$1.236_5 \times 10^{+5}$	$1.513_4 \times 10^{+4}$	$1.690_0 \times 10^{+4}$	$3.631_0 \times 10^{+3}$	$3.789_6 \times 10^{+3}$	$8.364_3 \times 10^{+2}$	$8.926_5 \times 10^{+2}$
8	$1.224_3 \times 10^{+5}$	$1.309_6 \times 10^{+5}$	$1.975_9 \times 10^{+4}$	$2.270_3 \times 10^{+4}$	$4.663_8 \times 10^{+3}$	$5.090_3 \times 10^{+3}$	$9.166_7 \times 10^{+2}$	$9.707_4 \times 10^{+2}$
9	$1.282_3 \times 10^{+5}$	$1.405_5 \times 10^{+5}$	$2.787_3 \times 10^{+4}$	$3.165_7 \times 10^{+4}$	$5.528_7 \times 10^{+3}$	$6.040_0 \times 10^{+3}$	$1.295_7 \times 10^{+3}$	$1.410_8 \times 10^{+3}$
10	$1.482_8 \times 10^{+5}$	$1.588_4 - 10^{+5}$	$4.663_3 \times 10^{+4}$	$5.423_0 \times 10^{+4}$	$8.787_1 \times 10^{+3}$	$9.944_4 \times 10^{+3}$	$1.396_4 \times 10^{+3}$	$1.565_2 \times 10^{+3}$

[a] $a = 2.00$, $\sigma = 0.300$. See Section 6.2. From [BFS2]. Here and in Tables VI–VIII, the superscripts for the lower bounds are the truncation index and the order of the intermediate problem, respectively.

TABLE V—*Continued*

ν	$b/a = 2.000$		$b/a = 4.000$		$b/a = 8.000$	
	$\lambda_\nu^{50,46}$	λ_ν^{50}	$\lambda_\nu^{50,46}$	λ_ν^{50}	$\lambda_\nu^{50,46}$	λ_ν^{50}
1	$4.477_4 \times 10^{-2}$	$4.669_7 \times 10^{-2}$	$2.708_3 \times 10^{-3}$	$2.877_7 \times 10^{-3}$	$1.664_2 \times 10^{-4}$	$1.794_9 \times 10^{-4}$
2	$1.732_9 \times 10^{0}$	$1.814_1 \times 10^{0}$	$1.058_5 \times 10^{-1}$	$1.125_9 \times 10^{-1}$	$6.527_7 \times 10^{-3}$	$7.040_6 \times 10^{-3}$
3	$1.357_3 \times 10^{+1}$	$1.428_7 \times 10^{+1}$	$8.314_1 \times 10^{-1}$	$8.874_9 \times 10^{-1}$	$5.122_5 \times 10^{-2}$	$5.531_4 \times 10^{-2}$
4	$3.246_5 \times 10^{+1}$	$3.406_9 \times 10^{+1}$	$3.204_5 \times 10^{0}$	$3.433_5 \times 10^{0}$	$1.971_7 \times 10^{-1}$	$2.131_9 \times 10^{-1}$
5	$5.235_2 \times 10^{+1}$	$5.531_7 \times 10^{+1}$	$8.779_5 \times 10^{0}$	$9.438_7 \times 10^{0}$	$5.404_5 \times 10^{-1}$	$5.851_8 \times 10^{-1}$
6	$5.864_6 \times 10^{+1}$	$6.322_0 \times 10^{+1}$	$1.955_6 \times 10^{+1}$	$2.109_7 \times 10^{+1}$	$1.210_0 \times 10^{0}$	$1.311_9 \times 10^{0}$
7	$1.138_4 \times 10^{+2}$	$1.255_6 \times 10^{+2}$	$3.064_7 \times 10^{-1}$	$3.183_8 \times 10^{+1}$	$2.367_6 \times 10^{0}$	$2.570_2 \times 10^{0}$
8	$1.457_9 \times 10^{+2}$	$1.546_0 \times 10^{+2}$	$3.608_6 \times 10^{+1}$	$3.790_4 \times 10^{+1}$	$4.207_4 \times 10^{0}$	$4.573_0 \times 10^{0}$
9	$2.149_9 \times 10^{+2}$	$2.422_0 \times 10^{+2}$	$3.926_5 \times 10^{+1}$	$4.208_6 \times 10^{+1}$	$6.955_3 \times 10^{0}$	$7.568_7 \times 10^{0}$
10	$3.234_7 \times 10^{+2}$	$3.447_8 \times 10^{+2}$	$4.796_9 \times 10^{+1}$	$5.156_5 \times 10^{+1}$	$1.084_1 \times 10^{+1}$	$1.184_0 \times 10^{+1}$

5. Rectangular Free Plates (Tables VI–VIII)

Differential equation and boundary conditions:

$$\Delta\,\Delta u = \lambda u \qquad \text{in} \quad \Omega$$

$$\left.\begin{aligned} \sigma\,\Delta u + (1-\sigma)\frac{\partial^2 u}{\partial n^2} &= 0 \\ \frac{\partial(\Delta u)}{\partial n} + (1-\sigma)\frac{d(\partial^2 u/\partial n \partial t)}{ds} &= 0 \end{aligned}\right\} \quad \text{on} \quad \partial\Omega$$

$$\Omega: \qquad -a/2 < x < a/2, \quad -b/2 < y < b/2$$

(Other notations same as above.)

TABLE VI

RECTANGULAR PLATE—FUNCTIONS ODD WITH RESPECT TO x AND ODD WITH RESPECT TO y[a]

ν	$b/a = 1.00$		$b/a = 1.25$		$b/a = 1.50$		$b/a = 2.00$		$b/a = 4.00$		$b/a = 8.00$	
	$\lambda_\nu^{50,44}$	λ_ν^{50}	$\lambda_\nu^{50,44}$	λ_ν^{50}	$\lambda_\nu^{50,44}$	λ_ν^{50}	$\lambda_\nu^{50,44}$	λ_ν^{50}	$\lambda_\nu^{50,44}$	λ_ν^{50}	$\lambda_\nu^{50,44}$	λ_ν^{50}
1	10.71_3	11.34_7	6.863_6	7.237_3	4.694_5	4.989_7	2.605_2	2.760_9	$.6186_9$	$.6643_7$	$.1468_9$	$.1631_8$
2	276.4_6	302.5_5	146.1_2	159.3_1	83.95_8	91.65_4	37.26_3	40.49_6	6.609_6	7.192_5	1.397_8	1.558_8
3	352.9_3	374.5_3	286.1_6	304.0_3	262.8_0	280.2_7	197.0_6	217.9_4	24.52_4	27.09_4	4.300_8	4.830_7
4	$1324._5$	$1465._3$	865.5_2	963.3_8	553.9_8	608.2_2	253.8_1	267.2_6	67.86_4	75.61_8	9.665_2	10.95_1
5	$2412._4$	$2631._0$	$1101._7$	$1197._4$	667.2_0	731.3_7	458.4_6	497.9_9	155.7_5	176.8_5	18.75_5	21.42_4
6	$2700._5$	$2884._0$	$2434._5$	$2646._3$	$1739._6$	$1948._4$	724.4_2	809.5_8	231.2_7	243.3_1	33.33_7	38.33_4
7	$4820._4$	$5342._5$	$2723._6$	$3025._7$	$2126._2$	$2345._7$	978.8_6	$1084._6$	281.6_8	298.4_2	55.66_9	64.35_6
8	$5093._4$	$5597._5$	$3889._5$	$4312._0$	$2465._6$	$2635._6$	$1911._4$	$2179._2$	318.6_9	368.8_7	88.44_0	102.7_3
9	$9946._5$	$1107_7.$	$4364._8$	$4803._4$	$3444._2$	$3745._4$	$1986._0$	$2257._3$	386.6_9	420.7_7	134.6_2	157.1_4
10	$1054_4.$	$1159_1.$	$7223._5$	$8049._3$	$4262._1$	$4759._6$	$2382._9$	$2570._3$	553.1_4	616.1_7	196.1_2	229.5_9

[a] $a = 2.00$, $\sigma = 0.33$. See Section 6.2. From [BFS1].

TABLE VII

SQUARE PLATE—FUNCTIONS ODD WITH RESPECT TO x, ODD WITH RESPECT TO y, AND ODD WITH RESPECT TO INTERCHANGE OF x AND y[a]

	$k^1 = 13, k^2 = 17$		$k^1 = 23, k^2 = 23$	
	$\lambda_\nu^{25,30}$	λ_ν^{25}	$\lambda_\nu^{50,46}$	λ_ν^{50}
1	10.81_4	11.34_7	10.89_2	11.34_5
2	355.0_7	374.7_1	358.4_9	373.9_8
3	$1345._2$	$1465._5$	$1363._7$	$1464._4$
4	$2704._7$	$2885._0$	$2742._1$	$2878._9$
5	$5123._1$	$5599._3$	$5210._0$	$5589._3$
6	$1064_9.$	$1159_3.$	$1081_6.$	$1157_4.$
7	$1141_9.$	$1271_6.$	$1167_1.$	$1269_6.$
8	$1531_4.$	$1668_5.$	$1554_8.$	$1665_3.$
9	$2532_1.$	$2867_5.$	$2605_0.$	$2859_1.$
10	$2975_6.$	$3310_5.$	$3073_0.$	$3304_0.$

[a] Side length 2; $\sigma = 0.300$. See Section 6.2. From [BFS1].

TABLE VIII

SQUARE PLATE—FUNCTIONS ODD WITH RESPECT TO x, ODD WITH RESPECT TO y, AND ODD WITH RESPECT TO INTERCHANGE OF x AND y[a]

	$k^1 = 13, k^2 = 17$		$k^1 = 23, k^2 = 23$	
ν	$\lambda_\nu^{25,30}$	λ_ν^{25}	$\lambda_\nu^{50,46}$	λ_ν^{50}
1	11.90_9	12.46_0	11.99_1	12.45_8
2	360.4_6	376.5_6	363.3_3	376.1_3
3	$1417._5$	$1529._0$	$1435._3$	$1528._0$
4	$2752._3$	$2883._4$	$2779._9$	$2880._1$
5	$5296._8$	$5718._6$	$5375._4$	$5709._4$
6	$1097_8.$	$1160_7.$	$1109_1.$	$1159_7.$
7	$1176_1.$	$1304_3.$	$1201_6.$	$1301_7.$
8	$1569_3.$	$1686_3.$	$1590_2.$	$1683_7.$
9	$2602_7.$	$2923_6.$	$2678_3.$	$2913_5.$
10	$3095_3.$	$3306_5.$	$3156_8.$	$3303_0.$

[a] Side length 2, $\sigma = 0.225$. See Section 6.2. From [BFS1].

6. Rhombical Membranes

For descriptions of problems, see Section 6.3. In Tables IX and X, *both* upper and lower bounds were obtained by intermediate problems of order 15.

TABLE IX

Rhombical Membrane, Unit Side Length, Fixed on All Edges, Functions Even with Respect to Both Diagonals[a]

ν	$\theta = 15°$		$\theta = 30°$		$\theta = 45°$		$\theta = 60°$		$\theta = 75°$	
	$\frac{\lambda_\nu^{-15}}{\pi^2}$	$\frac{\lambda_\nu^{+15}}{\pi^2}$	$\frac{\lambda_\nu^{-15}}{\pi^2}$	$\frac{\lambda_\nu^{+15}}{\pi^2}$	$\frac{\lambda_\nu^{-15}}{\pi^2}$	$\frac{\lambda_\nu^{+15}}{\pi^2}$	$\frac{\lambda_\nu^{-15}}{\pi^2}$	$\frac{\lambda_\nu^{-15}}{\pi^2}$	$\frac{\lambda_\nu^{-15}}{\pi^2}$	$\frac{\lambda_\nu^{-15}}{\pi^2}$
1	2.1137	2.1163	2.5210	2.5307	3.5170	3.5191	6.3150	6.4356	20.165	21.011
2	7.9960	8.0286	8.4807	8.5365	10.143	10.257	14.940	15.272	36.252	39.531
3	11.018	11.072	14.186	14.416	18.659	19.229	25.126	26.918	37.641*	71.754
4	17.036	17.225	17.076	17.340	22.039	22.631	37.108	45.457	40.963*	157.97
5	22.375	22.766	26.620	27.896	29.541	32.971	39.656*	51.153	41.710*	171.45
6	27.846	28.035	29.248	30.347	39.231	41.907	42.871*	78.736	45.779*	280.94
7	29.240	30.058	35.569	36.358	42.546	57.435	43.394	119.76	49.849*	457.81
8	37.282	38.842	39.866	47.788	43.348*	70.057	43.943*	138.00	50.866*	511.18
9	42.747	44.728	44.170	56.699	46.862*	79.159	48.230*	150.61	52.901*	546.25
10	44.572	49.894	49.333*	60.552	48.034*	95.197	52.518*	194.25	53.437	732.20

[a] See Section 6.3. Asterisks indicate persistent eigenvalues. From [S2].

TABLE X

RHOMBICAL MEMBRANE, UNIT SIDE LENGTH, FIXED ON TWO OPPOSITE EDGES AND FREE ON OTHER EDGES, FUNCTIONS EVEN WITH RESPECT TO MAIN DIAGONAL[a]

	$\theta = 15^\circ$		$\theta = 30^\circ$		$\theta = 45^\circ$		$\theta = 60^\circ$		$\theta = 75^\circ$	
ν	$\frac{\lambda_\nu^{-15}}{\pi^2}$	$\frac{\lambda_\nu^{+15}}{\pi^2}$	$\frac{\lambda_\nu^{-15}}{\pi^2}$	$\frac{\lambda_\nu^{+15}}{\pi^2}$	$\frac{\lambda_\nu^{-15}}{\pi^2}$	$\frac{\lambda_\nu^{+15}}{\pi^2}$	$\frac{\lambda_\nu^{-15}}{\pi^2}$	$\frac{\lambda_\nu^{+15}}{\pi^2}$	$\frac{\lambda_\nu^{-15}}{\pi^2}$	$\frac{\lambda_\nu^{+15}}{\pi^2}$
1	1.0301	1.0806	1.1820	1.3220	1.5474	1.9137	2.5046	3.6533	6.8038	13.043
2	4.6092	4.8526	4.6585	5.1547	5.2513	6.2262	6.9881	9.5382	14.233	27.428
3	5.5775	5.9283	6.8002	7.7875	9.3235	11.860	12.884	23.451	20.855*	82.839
4	9.1966	9.8215	9.8390	12.276	10.970	15.799	16.726	26.032	20.855*	87.796
5	11.638	12.519	12.217	13.185	15.710	19.025	21.014	38.595	22.889*	145.71
6	14.047	15.612	15.443	19.846	16.335	29.271	21.971*	57.303	22.889*	210.62
7	16.598	18.633	17.090	24.279	21.149	37.774	21.971*	77.301	24.195	292.15
8	18.096	19.553	21.454	26.374	24.017*	42.909	24.115*	91.324	24.924*	353.87
9	21.260	27.418	21.761	34.439	24.017*	51.994	24.115*	104.41	26.950*	3901.54
10	23.307	27.818	23.728	35.554	25.734	54.878	26.259*	112.72	26.950*	428.61

[a] See Section 6.3. Asterisks indicate persistent eigenvalues. From [S2].

7. Uniform Beam Rotating about a Simply Supported End
(Tables XI and XII)

Differential equation and boundary conditions:

$$u'''' - (a^2/2)[(1 - x^2)u']' = \lambda u \qquad (0 < x < 1),$$
$$u(0) = u''(0) = u''(1) = u'''(1) = 0.$$

TABLE XI[a]

	$a^2 = 5.0$		$a^2 = 50.0$		$a^2 = 500.0$	
ν	${}_S\lambda_\nu^{7,4}$	${}_S\lambda_\nu^{11}$	${}_S\lambda_\nu^{7,4}$	${}_S\lambda_\nu^{11}$	${}_S\lambda_\nu^{7,4}$	${}_S\lambda_\nu^{11}$
1	5.000000	5.000000	50.00000	50.00000	500.0000	500.0000
2	269.6703	269.6703	554.4686	554.4686	3316.361	3316.362
3	2585.934	2585.935	3384.199	3384.200	10958.27	10958.30
4	11046.94	11047.50	12655.42	12659.44	28158.83	28159.26
5	32050.80	32083.38	34496.11	34805.81	59447.73	61333.86
6	74291.20	74461.55	76907.85	78590.17	103385.1	119165.7

[a] See Section 6.4. Here, the superscripts 7, 4 for the lower bounds are the truncation indices and the superscript 11 for the upper bounds is the dimension of the space in the Rayleigh–Ritz method. From [RS1].

TABLE XII[a]

	$a^2 = 5.0$		$a^2 = 50.0$		$a^2 = 500.0$	
ν	${}_T\lambda_\nu^{15,15}$	${}_T\lambda_\nu^{15}$	${}_T\lambda_\nu^{15,15}$	${}_T\lambda_\nu^{15}$	${}_T\lambda_\nu^{15,15}$	${}_T\lambda_\nu^{15}$
1	5.000000	5.000000	50.00000	50.00000	500.0000	500.0000
2	268.6508	269.7044	544.8025	557.2948	3223.005	3525.917
3	2582.541	2585.991	33509.49	3390.092	10649.73	11353.97
4	11040.00	11047.56	12584.51	12665.50	27411.13	28703.15
5	32070.84	32083.41	34679.92	34811.71	60039.57	61924.23
6	74439.84	74460.44	78382.70	78595.18	116997.2	119763.5

[a] See Section 5.5. Here, the superscripts 15, 15, for the lower bounds are the truncation index and the order of the intermediate problem, while that of the upper bounds is as in Table XI.

8. Uniform Beam Rotating about a Fixed End (Table XIII)

Same differential equation as above, but with boundary conditions

$$u(0) = u'(0) = u''(1) = u'''(1) = 0.$$

TABLE XIII[a]

	$a^2 = 0$		$a^2 = 5$		$a^2 = 200$		$a^2 = 10{,}000$	
ν	$\lambda_\nu^{7,4}$	λ_ν^{7}	$\lambda_\nu^{7,4}$	λ_ν^{7}	$\lambda_\nu^{7,4}$	λ_ν^{7}	$\lambda_\nu^{7,4}$	λ_ν^{7}
1	12.362	12.362	18.287	18.301	231.74	233.80	10097	10297
2	485.52	485.52	517.41	517.91	1751.1	1771.7	60936	62004
3	3806.5	3808.8	3891.4	3897.9	7167.0	7305.5	157810	161150
4	14617	14670	14784	14849	21270	21812	312860	321650

[a] See Section 6.4. Here, the superscripts 7, 4 denote the truncation indices and 7 denotes the Rayleigh–Ritz dimension. From [BF9].

9. A Beam of Varying Stiffness (Table XIV)

Differential equation and boundary conditions:

$$[1 + a^2(\cos^2 x)u'']'' = \lambda u \qquad (-\pi/2 < x < \pi/2),$$

$$u(-\pi/2) = u''(-\pi/2) = u(\pi/2) = u''(\pi/2) = 0.$$

TABLE XIV

$a^2 = 2$, FUNCTIONS EVEN WITH RESPECT TO x[a]

	Base eigenvalues	Lower bounds $N=8$	Lower bounds $N=16$	Lower bounds $N=21$	Upper bounds
1	0	0.23639 (1)	0.23639 (1)	0.23639 (1)	0.23639 (1)
2	0.81 (2)	0.14965 (3)	0.14965 (3)	0.14965 (3)	0.14965 (3)
3	0.625 (3)	0.11490 (4)	0.11490 (4)	0.11490 (4)	0.11490 (4)
4	0.2401 (4)	0.44066 (4)	0.44067 (4)	0.44067 (4)	0.44067 (4)
5	0.6561 (4)	0.12024 (5)	0.12033 (5)	0.12033 (5)	0.12033 (5)
6	0.14641 (5)	0.26508 (5)	0.26842 (5)	0.26842 (5)	0.26842 (5)
7	0.28561 (5)	0.48106 (5)	0.52351 (5)	0.52351 (5)	0.52351 (5)
8	0.50625 (5)	0.83521*(5)	0.92781 (5)	0.92781 (5)	0.92781 (5)
9	0.83521 (5)		0.15306 (6)	0.15306 (6)	0.15306 (6)
10	0.13032 (6)		0.23880 (6)	0.23880 (6)	0.23880 (6)
11	0.19448 (6)		0.35631 (6)	0.35636 (6)	0.35636 (6)
12	0.27984 (6)		0.51194 (6)	0.51275 (6)	0.51275 (6)
13	0.39063 (6)		0.70624 (6)	0.71571 (6)	0.71571 (6)
14	0.53144 (6)		0.92835 (6)	0.97369 (6)	0.97370 (6)
15	0.70728 (6)		0.11859*(7)	0.12957 (7)	0.12959 (7)
16	0.92352 (6)		0.12510 (7)	0.16905 (7)	0.16925 (7)
17	0.11859 (7)			0.21573 (7)	0.21777 (7)
18	0.15006 (7)			0.26701 (7)	0.27835 (7)
19	0.18742 (7)			0.33062 (7)	0.35890 (7)
20	0.23134 (7)			0.34188*(7)	0.47549 (7)
21	0.28258 (7)			0.41006*(7)	0.66553 (7)

[a] See Section 5.4. Asterisks indicate persistent eigenvalues. All figures rounded to five significant digits. Notation: 0.23639 (1) = 0.2369×10^1. For the lower bounds, N denotes the number of base eigenvectors defining the special choice. For the upper bounds, the Rayleigh–Ritz dimension is 21. From [FR1].

10. A Beam of Linearly Varying Depth (Tables XV and XVI†)
Differential equation and boundary conditions:

$$[(1 + ax)^3 u'']'' = \lambda(1 + ax)u \qquad (0 < x < \pi),$$

$$u(0) = u''(0) = u(\pi) = u''(\pi) = 0.$$

TABLE XV

VALUES FOR $a\pi = 1$[a]

	Base eigenvalues	Lower bounds $\lambda^{15,12,8}$	Upper bounds
1	0.50000	2.0826	2.0826
2	8.0000	34.419	34.420
3	40.500	173.37	173.38
4	128.00	546.33	546.40
5	312.50	1331.4	1331.8

[a] See Sections 5.4, 5.5. For upper bounds, Rayleigh–Ritz dimension is 15. From [BF9].

TABLE XVI

VALUES FOR $a\pi = 1/19$[a]

	Lower bounds $\lambda^{25,20,15}$	Upper bounds
1	0.72156	0.72156
2	2.9334	2.9334
3	6.5837	6.5838
4	11.687	11.688
5	18.247	18.247
6	26.262	26.263
7	35.734	35.737
8	46.663	46.668
9	59.048	59.063
10	72.885	72.927

[a] See Sections 5.4, 5.5. Rayleigh–Ritz upper bounds are given. From [W24].

† In Tables XV and XVI, the superscripts give the truncation index and the numbers of eigenvectors defining the special choice for the numerator and for the denominator of the quadratic form, respectively.

11. A Clamped Beam (Table XVII)

Differential equation and boundary conditions:

$$u'''' = \lambda u \qquad (-\alpha\pi/2 < x < \alpha\pi/2),$$

$$u(-\alpha\pi/2) = u'(-\alpha\pi/2) = u(\alpha\pi/2) = u'(\alpha\pi/2) = 0.$$

TABLE XVII

$\alpha = 3/4$, FUNCTIONS EVEN WITH RESPECT TO x[a]

	Base eigenvalues	Lower bounds $\lambda^{15,15}$	Exact values
1	1.0000	14.158	16.241
2	81.000	413.36	474.28
3	625.00	2519.4	2892.0
4	2401.0	8704.2	10000.
5	6561.0	22377.	25742.

[a] See Sections 5.4, 5.5. The superscripts denote the truncation index and the number of special choice vectors, respectively. From [BF9].

12. The Schrödinger Equation for Helium (Table XVIII)

See Sections 5.7 and 5.8 for a detailed discussion of the problem. Differential equation:

$$-\tfrac{1}{2}\,\Delta_1 u - \tfrac{1}{2}\,\Delta_2 u - (2/r_1)u - (2/r_2)u + (1/r_{12})u = \lambda u$$
$$(-\infty < x_1, x_2, y_1, y_2, z_1, z_2 < \infty).$$

TABLE XVIII

HELIUM ATOM[a]

	Lower bounds (special choice)	Upper bounds
1	-3.063_7	-2.9037237
2	-2.165_5	-2.1458

[a] Lower bounds from [B1] and upper bounds from [K6].

13. **An Anharmonic Oscillator** (Table XIX)

Differential equation:

$$-u'' + x^2u + \varepsilon x^4 u = \lambda u \quad (-\infty < x < \infty).$$

TABLE XIX

FUNCTIONS EVEN WITH RESPECT TO x[a]

ε	λ_1	λ_2	λ_3	λ_4	λ_5
0.0	1.000000	5.000000	9.000000	13.000000	17.000000
	1.000000	5.000000	9.000000	13.000000	17.000000
0.1	1.065286	5.748178	11.10038	17.51524	30.94592
	1.065278	5.746596	10.95333	16.17279	21.00000*
0.2	1.118293	6.278820	12.48016	21.87339	45.99933
	1.118255	6.260404	12.22585	16.90845	21.00000*
0.3	1.164055	6.708557	13.67853	26.41021	61.16365
	1.163987	6.655885	13.25990	17.64313	21.00000*
0.4	1.204848	7.075869	14.82828	31.03013	76.36088
	1.204738	6.979830	14.03037	18.63119	21.00000*
0.5	1.241957	7.400376	15.96821	35.69220	91.57225
	1.241746	7.258083	14.55430	19.88068	21.00000*
0.6	1.276195	7.694107	17.11054	40.37815	106.7910
	1.275773	7.505763	14.90630	21.00000*	21.31832
0.7	1.308110	7.965074	18.25889	45.07885	122.0141
	1.307324	7.732038	15.15526	21.00000*	22.87282
0.8	1.338096	8.218847	19.41390	49.78925	137.2399
	1.336760	7.942661	15.34432	21.00000*	24.49895
0.9	1.366442	8.459408	20.57519	54.50637	152.4676
	1.364349	8.141353	15.49781	21.00000*	25.00000*
1.0	1.393371	8.689663	21.74203	59.22833	167.6966
	1.390301	8.330586	15.62953	21.00000*	25.00000*

[a] See Section 5.4. For each value of ε from 0 to 1.0, the Rayleigh–Ritz upper bounds are listed above and the lower bounds below. Asterisks indicate persistent eigenvalues. From [W24].

14. A Radial Schrödinger Equation (Tables XXa and XXb)

Differential equation:

$$-[xu]'' + [(x^2 + 1)/4x]u = \mu(1 - e^{-\alpha x/2k})u \qquad (0 < x < \infty).$$

TABLE XXa

LOWER BOUNDS[a]

l	k	$\mu_1^{l,k}$	$\mu_2^{l,k}$	$\mu_3^{l,k}$	$\mu_i^{l,k}$
1	0	1.000	2.000	2.000	2.000
1	1	1.234985	2.000	2.000	2.000
1	2	1.251726	2.000	2.000	2.000
2	0	1.000	2.000	3.000	3.000
2	1	1.248252	2.358910	3.000	3.000
2	2	1.258266	2.363134	3.000	3.000
3	0	1.000	2.000	3.000	4.000
3	1	1.249540	2.380424	3.349294	4.000
3	2	1.258725	2.394445	3.420742	4.000

[a] See Section 5.5. Here, l is the truncation index and k is the order of the intermediate problem. The condition on the last column is $i \geq 4$. From [BF1].

TABLE XXb

UPPER BOUNDS[a]

μ_1^{R4}	μ_2^{R4}	μ_3^{R4}	μ_4^{R4}
1.259005	2.416443	3.557628	6.091083

[a] See Section 2.3. Rayleigh–Ritz dimension is four. From [BF1].

15. **Mathieu's Equation** (Tables XXI–XXVII)

Differential equation and boundary conditions:

$$-u'' + s(\cos^2 x)u = \lambda u \qquad (0 < x < \pi),$$

$$u'(0) = u'(\pi) = 0$$

$$u(\pi/2 - x) = u(\pi/2 + x).$$

In Tables XXI–XXVII, k is the order of the intermediate problem and N is the truncation index.

TABLE XXI

EVEN SYMMETRY CLASS, $s = 1$, $k = 24$, $N = 25$[a]

Eigenvalue number	Lower bounds (special choice)	Upper bounds (Rayleigh–Ritz)	Upper bounds (special choice)
1	0.46896061E 00	0.46896061E 00	0.46896061E 00
2	0.45258291E 01	0.45258291E 01	0.45258291E 01
3	0.16502085E 02	0.16502085E 02	0.16502085E 02
4	0.36500894E 02	0.36500894E 02	0.36500893E 02
5	0.64500497E 02	0.64500497E 02	0.64500497E 02
6	0.10050032E 03	0.10050032E 03	0.10050032E 03
7	0.14450022E 03	0.14450022E 03	0.14450022E 03
8	0.19650016E 03	0.19650016E 03	0.19650016E 03
9	0.25650012E 03	0.25650012E 03	0.25650012E 03
10	0.32450010E 03	0.32450010E 03	0.32450010E 03
11	0.40050008E 03	0.40050009E 03	0.40050008E 03
12	0.48450007E 03	0.48450007E 03	0.48450007E 03
13	0.57650006E 03	0.57650007E 03	0.57650007E 03
14	0.67650000E 03	0.67650005E 03	0.67650005E 03
15	0.78450000E 03	0.78450005E 03	0.78450005E 03
16	0.90050004E 03	0.90050004E 03	0.90050004E 03
17	0.10245000E 04	0.10245000E 04	0.10245001E 04
18	0.11565000E 04	0.11565000E 04	0.11565000E 04
19	0.12965000E 04	0.12965000E 04	0.12965000E 04
20	0.14445000E 04	0.14445000E 04	0.14445000E 04
21	0.16005000E 04	0.16005000E 04	0.16005000E 04
22	0.17645000E 04	0.17645000E 04	0.17645000E 04
23	0.19365000E 04	0.19365000E 04	0.19365000E 04
24	0.21165000E 04	0.21165000E 04	0.21165000E 04
25	0.23042503E 04	0.23045003E 04	0.23047503E 04

[a] See Sections 5.4, 5.5. From [W24].

TABLE XXII

EVEN SYMMETRY CLASS, $s = 2$, $k = 24$, $N = 25$[a]

Eigenvalue number	Lower bounds (special choice)	Upper bounds (Rayleigh–Ritz)	Eigenvalue number	Lower bounds (special choice)	Upper bounds (Rayleigh–Ritz)
1	0.87823445E 00	0.87823447E 00	14	0.67700019E 03	0.67700019E 03
2	0.51009000E 01	0.51009006E 01	15	0.78500016E 03	0.78500017E 03
3	0.17008364E 02	0.17008364E 02	16	0.90100010E 03	0.90100014E 03
4	0.37003572E 02	0.37003572E 02	17	0.10250001E 04	0.10250001E 04
5	0.65001980E 02	0.65001984E 02	18	0.11570001E 04	0.11570001E 04
6	0.10100126E 03	0.10100126E 03	19	0.12970001E 04	0.12970001E 04
7	0.14500088E 03	0.14500088E 03	20	0.14450000E 04	0.14450000E 04
8	0.19700064E 03	0.19700064E 03	21	0.16010001E 04	0.16010001E 04
9	0.25700049E 03	0.25700049E 03	22	0.17650001E 04	0.17650001E 04
10	0.32500038E 03	0.32500039E 03	23	0.19370001E 04	0.19370001E 04
11	0.40100032E 03	0.40100032E 03	24	0.21170000E 04	0.21170001E 04
12	0.48500026E 03	0.48500026E 03	25	0.23045013E 04	0.23050013E 04
13	0.57700023E 03	0.57700023E 03			

[a] See Sections 5.4, 5.5. From [W24].

TABLE XXIII

EVEN SYMMETRY CLASS, $s = 4$, $k = 24$, $N = 25$[a]

Eigenvalue number	Lower bounds (special choice)	Upper bounds (Rayleigh–Ritz)	Eigenvalue number	Lower bounds (special choice)	Upper bounds (Rayleigh–Ritz)
1	0.15448613E 01	0.15448613E 01	14	0.67800070E 03	0.67800075E 03
2	0.63713010E 01	0.63713010E 01	15	0.78600064E 03	0.78600064E 03
3	0.18033832E 02	0.18033832E 02	16	0.90200055E 03	0.90200055E 03
4	0.38014290E 02	0.38014290E 02	17	0.10260005E 04	0.10260005E 04
5	0.66007938E 02	0.66007938E 02	18	0.11580000E 04	0.11580004E 04
6	0.10200505E 03	0.10200505E 03	19	0.12980004E 04	0.12980004E 04
7	0.14600350E 03	0.14600350E 03	20	0.14460003E 04	0.14460003E 04
8	0.19800256E 03	0.19800257E 03	21	0.16020003E 04	0.16020003E 04
9	0.25800196E 03	0.25800196E 03	22	0.17660002E 04	0.17660002E 04
10	0.32600155E 03	0.32600155E 03	23	0.13980002E 04	0.19380002E 04
11	0.40200125E 03	0.40200125E 03	24	0.21180002E 04	0.21180002E 04
12	0.48600104E 03	0.48600104E 03	25	0.23050053E 04	0.23060053E 04
13	0.57800087E 03	0.57800087E 03			

[a] See Sections 5.4, 5.5. From [W24].

TABLE XXIV

EVEN SYMMETRY CLASS, $s = 8$, $k = 24$, $N = 25$[a]

Eigenvalue number	Lower bounds (special choice)	Upper bounds (Rayleigh–Ritz)	Eigenvalue number	Lower bounds (special choice)	Upper bounds (Rayleigh–Ritz)
1	0.24860431E 01	0.24860431E 01	14	0.68000297E 03	0.68000297E 03
2	0.91726652E 01	0.91726652E 01	15	0.78800250E 03	0.78800256E 03
3	0.20141204E 02	0.20141204E 02	16	0.90400223E 03	0.90400223E 03
4	0.40057216E 02	0.40057216E 02	17	0.10280019E 04	0.10280019E 04
5	0.68031758E 02	0.68031758E 02	18	0.11600017E 04	0.11600018E 04
6	0.10402020E 03	0.10402020E 03	19	0.13000015E 04	0.13000015E 04
7	0.14801399E 03	0.14801399E 03	20	0.14480013E 04	0.14480013E 04
8	0.20001025E 03	0.20001025E 03	21	0.16040012E 04	0.16040012E 04
9	0.26000785E 03	0.26000785E 03	22	0.17680010E 04	0.17680011E 04
10	0.32800619E 03	0.32800619E 03	23	0.19400010E 04	0.19400010E 04
11	0.40400500E 03	0.40400501E 03	24	0.21200007E 04	0.21200009E 04
12	0.48800415E 03	0.48800514E 03	25	0.23060215E 04	0.23080213E 04
13	0.58000340E 03	0.58000348E 03			

[a] See Sections 5.4, 5.5. From [W24].

TABLE XXV

EVEN SYMMETRY CLASS, $s = 16$, $k = 24$, $N = 25$[a]

Eigenvalue number	Lower bounds (special choice)	Upper bounds (Rayleigh–Ritz)	Eigenvalue number	Lower bounds (special choice)	Upper bounds (Rayleigh–Ritz)
1	0.37194810E 01	0.37194811E 01	14	0.68401185E 03	0.68401186E 03
2	0.14829075E 02	0.14829075E 02	15	0.79201024E 03	0.79201024E 03
3	0.24649819E 02	0.24649819E 02	16	0.90800890E 03	0.90800890E 03
4	0.44229950E 02	0.44229952E 02	17	0.13020078E 04	0.10320078E 04
5	0.72127160E 02	0.72127160E 02	18	0.11640069E 04	0.11640069E 04
6	0.10808085E 03	0.10808085E 03	19	0.13040061E 04	0.13040061E 04
7	0.15205596E 03	0.15205596E 03	20	0.14520055E 04	0.14520055E 04
8	0.20404103E 03	0.20404103E 03	21	0.16080049E 04	0.16080050E 04
9	0.26403137E 03	0.26403137E 03	22	0.17720045E 04	0.17720045E 04
10	0.33202477E 03	0.33202477E 03	23	0.19440041E 04	0.19440041E 04
11	0.40802005E 03	0.40802006E 03	24	0.21240020E 04	0.21240038E 04
12	0.49201657E 03	0.49201657E 03	25	0.23080869E 04	0.23120851E 04
13	0.58401392E 03	0.58401392E 03			

[a] See Sections 5.4, 5.6. From [W24].

TABLE XXVI

EVEN SYMMETRY CLASS, $s = 32$, $k = 24$, $N = 25$[a]

Eigenvalue number	Lower bounds (special choice)	Upper bounds (Rayleigh–Ritz)	Eigenvalue number	Lower bounds (special choice)	Upper bounds (Rayleigh–Ritz)
1	0.53932708E 01	0.53932708E 01	14	0.69204742E 03	0.69204742E 03
2	0.24115238E 02	0.24115239E 02	15	0.80004088E 03	0.80004088E 03
3	0.35252705E 02	0.35252706E 02	16	0.91603561E 03	0.91603561E 03
4	0.52949087E 02	0.52949087E 02	17	0.10400312E 04	0.10400312E 04
5	0.80510818E 02	0.80510818E 02	18	0.11720277E 04	0.11720277E 04
6	0.11632393E 03	0.11632393E 03	19	0.13120247E 04	0.13120247E 04
7	0.16022400E 03	0.16022400E 03	20	0.14600222E 04	0.14600222E 04
8	0.21216419E 03	0.21216419E 03	21	0.16160200E 04	0.16160200E 04
9	0.27212553E 03	0.27212553E 03	22	0.17800181E 04	0.17800181E 04
10	0.34009909E 03	0.34009909E 03	23	0.19520165E 04	0.19520165E 04
11	0.41608021E 03	0.41608021E 03	24	0.21320003E 04	0.21320154E 04
12	0.50006627E 03	0.50006627E 03	25	0.23123552E 04	0.23203401E 04
13	0.59205566E 03	0.59205566E 03			

[a] See Sections 5.4, 5.5. From [W24].

TABLE XXVII

EVEN SYMMETRY CLASS, $s = 1000$, $k = 25$[a]

μ	Lower bounds (special choice)	Upper bounds	
		Rayleigh–Ritz	Special choice
1	0.31371 (2)	0.31371 (2)	0.31371 (2)
2	0.15479 (3)	0.15479 (3)	0.15479 (3)
3	0.27394 (3)	0.27394 (3)	0.27394 (3)
4	0.38856 (3)	0.38856 (3)	0.38856 (3)
5	0.49834 (3)	0.49834 (3)	0.49834 (3)
6	0.60285 (3)	0.60285 (3)	0.60285 (3)
7	0.70157 (3)	0.70157 (3)	0.70157 (3)
8	0.79370 (3)	0.79370 (3)	0.79370 (3)
9	0.87798 (3)	0.87798 (3)	0.87798 (3)
10	0.95098 (3)	0.95098 (3)	0.95098 (3)
11	0.10025 (4)	0.10025 (4)	0.10025 (4)
12	0.10572 (4)	0.10572 (4)	0.10572 (4)
13	0.11344 (4)	0.11344 (4)	0.11344 (4)
14	0.12246 (4)	0.12246 (4)	0.12246 (4)
15	0.13253 (4)	0.13253 (4)	0.13253 (4)
16	0.14357 (4)	0.14357 (4)	0.14357 (4)
17	0.15552 (4)	0.15552 (4)	0.15552 (4)
18	0.16835 (4)	0.16835 (4)	0.16835 (4)
19	0.18204 (4)	0.18204 (4)	0.18205 (4)
20	0.19653 (4)	0.19660 (4)	0.19663 (4)
21	0.21157 (4)	0.21207 (4)	0.21227 (4)
22	0.22639 (4)	0.22879 (4)	0.22962 (4)
23	0.24219 (4)	0.24778 (4)	0.25014 (4)
24	0.25000*(4)	0.27101 (4)	0.27654 (4)
25	0.26334 (4)	0.30247 (4)	0.31851 (4)

[a] See Section 5.4. From [FR1]. Asterisk indicates persistent eigenvalue. All figures rounded to five significant digits. 0.31371 (2) = 0.31371×10^2.

Bibliography

Books and Monographs

1. Beckenbach, E. F., and Bellman, R., *Inequalities*. Springer, Berlin, 1961.
2. Bellman, R., *Introduction to Matrix Analysis*. McGraw-Hill, New York, 1960.
3. Courant, R., and Hilbert, D., *Methods of Mathematical Physics*, Vols. I and II. Wiley (Interscience), New York, 1953 and 1965.
4. Dunford, N., and Schwartz, J. T., *Linear Operators*. Part I: General Theory, Part II: Spectral Theory. Wiley (Interscience), New York, 1958 and 1963.
5. Fichera, G., *Transformazioni Lineari*, 3rd ed. Veschi, Rome, 1962.
6. Fichera, G., *Linear Elliptic Differential Systems and Eigenvalue Problems* (Lecture Notes in Mathematics). Springer, New York, 1965.
7. Friedrichs, K. O., *Spectral Theory of Operators in Hilbert Space*. Institute of Mathematical Sciences, New York University, 1960.
8. Friedrichs, K. O., *Perturbation of Spectra in Hilbert Space* (Lectures in Applied Mathematics). Amer. Math. Soc., Providence, Rhode Island, 1965.
9. Gantmakher, F. R., *Theory of Matrices*. Chelsea, New York, 1959.
10. Gohberg, I. C., and Krein, M. G., *Introduction to the Theory of Linear Nonselfadjoint Operators* (Russian). 1965. [Transl.: Math. Monographs, Amer. Math. Soc., Providence, Rhode Island, 1969.]
11. Goldberg, S., *Unbounded Linear Operators: Theory and Applications*. McGraw-Hill, New York, 1966.
12. Gould, S. H., *Variational Methods for Eigenvalue Problems: An Introduction to the Weinstein Method of Intermediate Problems*, 2nd ed. Univ. of Toronto Press, 1966.

12a. Gould, S. H., *Variational Methods for Eigenvalue Problems: An Introduction to the Weinstein Method of Intermediate Problems*, (Russian translation, V. B. Lidskii, ed.). Moscow, 1970.

13. Halmos, P. R., *Finite Dimensional Vector Spaces*. Princeton Univ. Press, Princeton, New Jersey, 1942.
14. Halmos, P. R., *Introduction to Hilbert Space and the Theory of Spectral Multiplicity*. Chelsea, New York, 1951.

15. HALMOS, P. R., *A Hilbert Space Problem Book*. Van Nostrand-Reinhold, Princeton, New Jersey, 1967.
16. HAMBURGER, H., and GRIMSHAW, M. E., *Linear Transformations in n-dimensional Vector Space*. Cambridge Univ. Press, New York and London, 1951.
17. HARDY, G. H., LITTLEWOOD, J. E., and PÓLYA, G., *Inequalities*, 2nd ed., Cambridge Univer. Press, New York and London, 1952.
18. HELLWIG, G., *Partial Differential Equations*. Ginn (Blaisdell), Boston, Massachusetts, 1964.
19. KATO, T., *Perturbation Theory for Linear Operators*. Springer, New York, 1966.
20. KEMBLE, E. C., *The Fundamental Principles of Quantum Mechanics*. Dover, New York, 1958.
21. MIKHLIN, S. G., *Integral Equations*. Macmillan, New York, 1964.
22. MIKHLIN, S. G., *Variational Methods in Mathematical Physics*. Macmillan, New York, 1964 (2nd. Russian edition 1970).
23. NOODLEMAN, J. L., *Methods for Eigenvalues of Beams* (Russian). Modern Problems of Mechanics, Moscow, 1949.
24. PÓLYA, G., and SZEGÖ, G., *Isoperimetric Inequalities in Mathematical Physics*. Princeton Univ. Press, Princeton, New Jersey, 1951.
25. RAYLEIGH, LORD (STRUTT, J. W.), *The Theory of Sound*, 2nd ed., Dover, New York, 1945.
26. RIESZ, F., and SZ.-NAGY, B., *Functional Analysis*. Ungar, New York, 1955.
27. RITZ, W., *Oeuvres*. Gauthier-Villars, Paris, 1911.
28. STONE, M. H., *Linear Transformations in Hilbert Space*. Amer. Math. Soc. Colloquium Publications, Vol. 15, New York, 1932.
29. SZ.-NAGY, B., *Extensions of Linear Transformations in Hilbert Space which Extend Beyond this Space* (Appendix to Functional Analysis). Ungar, New York, 1960.
30. TEMPLE, G., and BICKLEY, W. G., *Rayleigh's Principle and its Applications to Engineering*. Constable, London, 1933.
31. FLÜGGE, W., *Handbook of Engineering Mechanics* (Sec. 4.1). McGraw-Hill, New York, 1962.
32. SMITHIES, F., *Integral Equations*. Cambridge Univ. Press, London, 1958.
33. WEINBERGER, H. F., *Variational Methods for Eigenvalue Problems*. Univ. of Minnesota Press, Minneapolis, 1962.
34. WEINBERGER, H. F., *Variational Methods in Boundary Value Problems*. Univ. of Minnesota Press, Minneapolis, 1961.

Articles

ARONSZAJN, N.

A1. The Rayleigh–Ritz and A. Weinstein methods for approximation of eigenvalues, I, II, *Proc. Nat. Acad. Sci., U.S.A.* **34**, 474–480, 594–601 (1948).

A2. The Rayleigh–Ritz and A. Weinstein methods for approximation of eigenvalues. Tech. Rept. 1, Oklahoma State U. of Agric. and Applied Science, Stillwater, Oklahoma, 1949.

A3. Introduction to the theory of Hilbert spaces, Research Foundation Report, Stillwater, Oklahoma, 1950.

A4. Theory of reproducing kernels, *Trans. Amer. Math. Soc.* **68**, 337 (1950).

A5. Approximation methods for eigenvalues of completely continuous symmetric operators, *Proc. Symp. Spectral Theory and Differential Problems*, pp. 179–202. Stillwater, Oklahoma, 1951.

A6. Green's functions and reproducing kernels, *Proc. Symp. Spectral Theory and Differential Problems*, pp. 355–412. Stillwater, Oklahoma, 1951.

A7. Application of Weinstein's Method with an Auxiliary Problem of Type I. Office Nav. Res. Tech. Rept., Univ. of Kansas, Lawrence, Kansas, 1951.

A8. Application of Weinstein's Method with an Auxiliary Problem of Type II, Office Nav. Res. Tech. Rept., Univ. of Kansas, Lawrence, Kansas, 1952.

A9. Differential Operators. Office Nav. Res. Tech. Rept., Univ. of Kansas, Lawrence, Kansas, 1953.

A10. Operators in a Hilbert Space. Office of Naval Research Tech. Rept., Univ. of Kansas, Lawrence, Kansas, 1953.

ARONSZAJN, N., and DONOGHUE, W. F.

AD1. Variational Approximation Methods Applied to Eigenvalues of a Clamped Rectangular Plate, Part I, Auxiliary Problems, studies in Eigenvalue Problems. Tech. Rept. 12, Department of Mathematics, Univ. of Kansas, Lawrence, Kansas, 1954.

ARONSZAJN, N., and MILGRAM, A. N.

AM1. Differential Operators on Riemannian Manifolds. Office of Naval Research Tech. Rept. No. 8., Univ. of Kansas, Lawrence, Kansas, 1953.

ARONSZAJN, N., and SMITH, K. T.

AS1. Functional Spaces and Functional Completion. Office of Naval Research Tech. Report No. 10., Univ. of Kansas, Lawrence, Kansas, 1954.

AS2. Invariant subspaces of completely continuous operators. *Ann. of Math.* **60**, 345 (1954).

ARONSZAJN, N., and WEINSTEIN, A.

AW1. Existence, convergence, and equivalence in the unified theory of eigenvalues of plates and membranes, *Proc. Nat. Acad. Sci. U.S.A.* **27**, 188–191 (1941).

AW2. On the unified theory of eigenvalues of plates and membranes, *Amer. J. Math.* **64**, 623–645 (1942).

ARONSZAJN, N., and ZEICHNER, A.

AZ1. Preliminary Note: Reproducing and Pseudo-Reproducing Kernels and Their Application to the Partial Differential Equations of Physics, Office of Naval Research Tech. Rept., Univ. of Kansas, Lawrence, Kansas, 1954.

BAZLEY, N. W.

B1. Lower bounds for eigenvalues with application to the helium atom, *Proc. Nat. Acad. Sci. U.S.A.* **45**, 850–853 (1959).

B2. Lower bounds for eigenvalues with application to the helium atom, *Phys. Rev.* **129**, 144–149 (1960).

B3. Lower bounds for eigenvalues, *J. Math. Mech.* **10**, 289–308 (1961).

BAZLEY, N. W., BÖRSCH-SUPAN, W., and FOX, D. W.

BBF1. Lower bounds for eigenvalues of a quadratic form relative to a positive quadratic form, *Arch. Rational Mech. Anal.* **27**, 398–406 (1968).

BAZLEY, N. W., and FOX, D. W.

BF1. Lower bounds for eigenvalues of Schrödinger's equation, *Phys. Rev.* **124**, 483–492 (1961).

BF2. Truncations in the method of intermediate problems for lower bounds to eigenvalues, *J. Res. Nat. Bur. Standards* Sect. B. **65**, 105–111 (1961).
BF3. Error bounds for eigenvectors of self-adjoint operators, *J. Res. Nat. Bur. Standards* Sect. B. **66**, 1–4 (1962).
BF4. A procedure for estimating eigenvalues, *J. Math. and Phys.* **3**, 469–471 (1962).
BF5. Lower bounds to eigenvalues using operator decompositions of the form B^*B, *Arch. Rational Mech. Anal.* **10**, 352–360 (1962).
BF6. Error bounds for expectation values, *Rev. Modern Phys.* **35**, 712–715 (1963).
BF7. Lower bounds for energy levels of molecular systems, *J. Math. and Phys.* **4**, 1147–1153 (1963).
BF8. Improvement of bounds to eigenvalues of operators of the form T^*T, *J. Res. Nat. Bur. Standards* Sect. B. **68**, 173–183 (1964).
BF9. Methods for lower bounds to frequencies of continuous elastic systems, *Z. Angew. Math. Phys.* **17**, 1–37 (1966).
BF10. Error bounds for approximations to expectation values of unbounded operators, *J. Math. and Phys.* **7**, 413–416 (1966).
BF11. Comparison operators for lower bounds to eigenvalues, *J. Reine Angew. Math.* **223**, 142–149 (1966).

BAZLEY, N. W., FOX, D. W., and STADTER, J. T.
BFS1. Upper and lower bounds for the frequencies of rectangular free plates, *Z. Angew. Math. Phys.* **18**, 445–460 (1967).
BFS2. Upper and lower bounds for the frequencies of rectangular clamped plates, *Z. Angew. Math. Mech.* **47**, 191–198 (1967).
BFS3. Upper and lower bounds for the frequencies of rectangular cantilever plates, *Z. Angew. Math. Mech.* **47**, 251–260 (1967).

BLANCH, G.
B4. Numerical aspects of Mathieu eigenvalues, *Rend. Circ. Mat. Palermo* **15**, 51–97 (1966).

BUDIANSKY, B., and HU, C. P.
BH1. The Lagrangian multiplier method of finding upper and lower limits to critical stresses of clamped plates. *Tech. Notes Nat. Advis. Comm. Aeronaut.*, 1103, Washington, D.C., 1946.

COLAUTTI, M. P.
C1. Su un teorema di completezza connesso al metodo di Weinstein per il calcolo degli autovalori, *Atti Accad. Sci. Torino Cl. Sci. Fis. Mat. Natur.* **97**, 1–21 (1962).

COOLIDGE, A., and JAMES, M.
CJ1. Wave functions for 1S 2S helium, *Phys. Rev.*, **49**, 676 (1936).

COURANT, R.
C2. Über die Eigenwerte bei den Differentialgleichungen der mathematischen Physik, *Math.* **7**, 1–57 (1920).

DAVIS, C.
D1. Compressions to finite-dimensional subspaces, *Proc. Amer. Math. Soc.* **9**, 356–359 (1958).

DE VITO, L., FICHERA, G., FUSCIARDI, A., and SCHAERF, M.
DFFS1. Sul calcolo degli autoralori della piastra quadrata incastrata lungo il bordo, *Rend. Accad. Naz. Lincei.* **40**, 725–733 (1966).

DIAZ, J. B.

D2. Upper and lower bounds for eigenvalues, *Proc. Eighth Symp. Appl. Math., Amer. Math. Soc.*, McGraw-Hill, New York, 1958.

DIAZ, J. B., and METCALF, F. T.

DM1. A functional equation for the Rayleigh quotient for eigenvalues and some applications, *J. Math. Mech.* **17**, 623–630 (1968).

DOLBERG, M. D.

D3. On links of greatest rigidity (Russian), *Ucen. Zap. Harkovsk. Gos. Univ.* **25,** 179–190 (1957).

FAN, K.

F1. On a theorem of Weyl concerning eigenvalues of linear transformations, I, *Proc. Nat. Acad. Sci.* **35**, 652–655 (1949).

FICHERA, G.

F2. Sul calcolo degli autovalori, *Congresso Soc. Ital. Progr. Sci.* pp. 1–19. Rome, Italy, 1965.

F3. Approximation and estimates for eigenvalues, *Proc. Symp. Numerical Solution of Partial Differential Equations*, Academic Press, New York, 1966.

F4. On the compactness of the base operator in the theory of intermediate problems, *Rend. Accad. Naz. Linei.* **41**, 3–7 (1966).

F5. Generalized biharmonic problem and related eigenvalue problems, *Blanch Anniversary Volume*, pp. 35–44. U.S.A.F., 1967.

F6. Sul miglioramento delle approssimazioni per difetto degli autovalori, I e II, *Rend. Accad. Naz. Lincei.* **42**, 138–145, 331–340 (1967).

F7. Il calcolo degli autovalori, *Boll. Un. Mat. Ital.* **1**, 33–95 (1968).

FISCHER, E.

F8. Über quadratische Formen mit reellen Koeffizienten, *Monatsh. Math. Phys.* **16**, 234–249 (1905).

FOX, D. W., and RHEINBOLDT, W. C.

FR1. Computational methods for determining lower bounds for eigenvalues of operators in Hilbert space, *SIAM Rev.* **8**, 427–462 (1966).

FUJITA, H.

F9. Perturbation by degenerate operators and Weinstein's method (Japanese), *Nippon Butsuri Gakkaisi.* **13**, 364–370 (1958).

GÅRDING, L.

G1. On a lemma by H. Weyl, *Fysiografiska Sällskapets i Lund Förhandlinger.* **20,** 1–4 (1950).

GAY, J. G.

G2. Lower Bounds to the Eigenvalues of Hamiltonians by Intermediate Problems. Doctoral dissertation, Univ. of Florida, Gainesville, Florida, 1963.

G3. *Phys. Rev.* **135A**, A 1220 (1964).

GOHBERG, I. C., and KREIN, M. G.

GK1. Fundamental theorems on deficiency numbers, root numbers, and indices of linear operators (Russian), *Uspehi Mat. Nauk* **12**, 43–118 (1957). (Translated in *Amer. Math. Soc. Transls.*, Ser. 2, Vol. 13.)

GUSTAFSON, K.

G5. On projections of self-adjoint operators and operator product adjoints, *Bull. Amer. Math. Soc.* **75**, 739–741 (1969).

HADAMARD, J.

H1. Newton and the infinitesimal calculus, *Newton Tercentenary Celebrations*, p. 35. The Royal Society, 1946.

HALMOS, P. R.

H2. Normal dilation and extensions of operators, *Summa Brasil. Math.* II. **9**, 1–10 (1950).

HERSCH, J.

H3. Caractérisation variationnelle d'une somme de valeurs propres consécutives; généralisation d'inégalités de Pólya-Schiffer et de Weyl, *Compt. Rend.* **252**, 1714–1716 (1961).

H4. Inégalités pour des valeurs propres consécutives de systémes vibrants inhomogénes, allant (en sens inverse) de celles de Pólya-Schiffer et de Weyl, *Compt. Rend.* **252**, 2496–2498 (1961).

HOLLAND, S.

H5. On the adjoint of the product of operators, *Bull. Amer. Math. Soc.* **74**, 931–932 (1968).

H6. On the adjoint of the product of operators, *J. Functional Analysis* **3**, 337–344 (1969).

HOOKER, W., and PROTTER, M. H.

HP1. Bounds for the first eigenvalue of a rhombic membrane, *J. Math. Phys.* **39**, 18–34, (1960).

HOWLAND, J. S.

H7. On the Weinstein–Aronszajn formula, *Arch. Rational Mech. Anal.* **39**, 323–339 (1970).

JAMES, H. M.

J1. Some applications of the Rayleigh–Ritz method to the theory of the structure of matter, *Bull. Amer. Math. Soc.* **47**, 869 (1941).

JENNINGS, A. K.

J2. Operators in Reproducing Kernel Spaces. Office of Naval Research Tech. Rept., Univ. of Kansas, Lawrence, Kansas, 1954.

J3. Some Developments and Applications of a New Approximation Method for Partial Differential Eigenvalue Problems. Office of Naval Research Tech. Rept., Univ. of Kansas, Lawrence, Kansas, 1951.

KATO, T.

K1. On the upper and lower bounds of eigenvalues, *J. Phys. Soc. Japan.* **4**, 334–339 (1949).

K2. Fundamental properties of Hamiltonian operators of Schrödinger type, *Trans. Amer. Math. Soc.* **70**, 195–211 (1951).

K3. On the existence of solutions of the helium wave equation, *Trans. Amer. Math. Soc.* **70**, 212–218 (1951).

K4. On some approximate methods concerning the operators T^*T, *Math. Ann.*, **126**, 253–262 (1953).

K5. Quadratic Forms in Hilbert Spaces and Asymptotic Perturbation Series. Tech. Rept. No. 7, Dept. of Math., Univ. of Calif., Berkeley, 1955.

KATO, T., FUJITA, H., NAKATA, Y., and NEWMAN, M.

KFNN1. Estimation of the frequencies of thin elastic plates with free edges, *J. Res. Nat. Bur. Standards Sect B.* **59**, 169–186 (1957).

KINOSHITA, T.

K6. Ground state of the helium atom, *Phys. Rev.* **105**, 1490 (1957).

KREIN, M. G.

K7. On the trace formula in perturbation theory (Russian), *Mat. Sb.* **33**, 597–626 (1953).

K8. On criteria of completeness of root vectors of dissipative operators (Russian), *Uspehi Mat. Nauk.* **3** (87), 145–152 (1959).

K9. On the theory of linear non-selfadjoint operators (Russian), *Dokl. Akad. Nauk SSSR* **2**, 254–256 (1960).

KURODA, S. T.

K10. On a generalization of the Weinstein-Aronszajn formula and the infinite determinant, *Sci. Papers College Gen. Ed. Univ. Tokyo.* **II**, 1–12 (1961).

K11. Finite dimensional perturbation and a representation of scattering operator, *Pacific J. Math.* **13**, 1305–1318 (1963).

K12. Stationary methods in the theory of scattering, *Perturbation Theory and its Applications in Quantum Mechanics* (C. H. Wilcox, ed.), 185–214, Wiley, New York, 1966.

LAX, P. D.

L1. A procedure for obtaining upper bounds for the eigenvalues of a hermitian symmetric operator, *Studies in Mathematical Analysis and Related Topics*, pp. 199–201. Stanford Univ. Press, Stanford, California, 1962.

LÖWDIN, P. O.

L2. Lower Bounds to Energy Eigenvalues. Quantum Theory Project Rept. Univ. of Florida, Gainesville, Florida, 1964.

L3. The calculation of upper and lower bounds of energy eigenvalues in perturbation theory by means of partitioning technique, *Perturbation Theory and its Applications in Quantum Mechanics* (C. H. Wilcox, ed.), pp. 255–294. Wiley, New York, 1966.

L4. Studies in perturbation theory, XIII. Treatment of constants of motion in resolvent method, partitioning technique, and perturbation theory, *Internat. J. Quantum Chem.* **2**, 867–931 (1968).

L5. Some properties of inner projections, *Internat. J. Quantum Chem.* **4**, 231–237 (1971).

MURRAY, F. J., and VON NEUMANN, J.

MN1. On rings of operators, *Ann. of Math.* **37**, 120–214 (1936).

NAKATA, Y., and FUJITA, H.

NF1. On upper and lower bounds of the eigenvalues of a free plate, *J. Phys. Soc. Japan.* **10**, 823–824 (1953).

PAYNE, L. E.

P1. Inequalities for eigenvalues of membranes and plates, *J. Rational Mech. Anal.* **4**, 517–529 (1955).

PAYNE, L. E., and WEINBERGER, H. F.

PW1. Lower bounds for vibration frequencies of elastically supported membranes and plates, *SIAM J. Appl. Math.* **5**, 171–182 (1957).

PW2. Some isoperimetric inequalities for membrane frequencies and torsional rigidity, *J. Math. Anal. Appl.* **2**, 210–216 (1961).

PLEIJEL, A.

P2. Propriétés asymptotiques des fonctions et valeurs propres de certains problèmes de vibration, *Ark. Mat., Astronom., Fys.* **27A**, 1–100 (1940).

P3. On Green's functions for elastic plates with clamped, supported and free edges, *Proc. Symp. Spectral Theory and Differential Problems*, pp. 413–483. Stillwater, Oklahoma, 1951.

POINCARÉ, H.

P4. Sur les équations aux dérivées partielles de la physique mathématique, *Amer. J. Math.* **12**, 211–294 (1890).

PÓLYA, G.

P5. Estimates for eigenvalues, *Studies in Mathematics and Mechanics*, presented to R. von Mises, pp. 200–207. Academic Press, New York, 1954.

PÓLYA, G., and SCHIFFER, M.

PS1. Convexity of functionals by transplantation, *J. Analyse Math.* **3**, 245–345 (1954).

RITZ, W.

R1. Über eine neue Methode zur Lösung gewisser Variations-probleme der mathematischen Physik, *J. Reine Angew. Math.* **135** (1908).

R2. Theorie der Transversalschwingungen einer quadratischen Platte mit freien Rändern, *Ann. Physik.* **38** (1909).

RUBINSTEIN, N., SIGILLITO, V. G., and STADTER, J. T.

RSS1. Bounds to Bending Frequencies of Non-Uniform Shafts, and Applications to Missiles. Tech. Rept. TG-966, Applied Physical Laboratory, Johns Hopkins Univ., 1968.

RUBINSTEIN, N., and STADTER, J. T.

RS1. Bounds to Frequencies of a Simply Supported Rotating Beam. Tech. Rept. TG-1042, Applied Physics Laboratory, Johns Hopkins Univ., 1968.

SERRIN, J.

S1. On the stability of viscous fluid motions, *Arch. Rational Mech. Annal.* **3**, 1 (1959).

STADTER, J. T.

S2. Bounds to eigenvalues of rhombical membranes, *SIAM J. Appl. Math.* **14** 324–341 (1966).

S3. Bounds for Eigenvalues of an Integral Operator. Tech. Rept. TG-979, Applied Physics Laboratory, Johns Hopkins Univ., 1968.

STAKGOLD, I.

S4. On Weinstein's intermediate problems for integral equations with difference kernels, *J. Math. Mech.* **19**, 301–308 (1969).

STENGER, W.

S5. On Poincaré's bounds for higher eigenvalues, *Bull. Amer. Math. Soc.* **72**, 715–718 (1966).

S6. A necessary and sufficient condition in Poincaré's theory of eigenvalues, *Notices Amer. Math. Soc.* **13**, 591 (1966).

S7. An inequality for the eigenvalues of a class of self-adjoint operators, *Bull. Amer. Math. Soc.* **73**, 487–490 (1967).

S8. On the variational principles for eigenvalues for a class of unbounded operators, *J. Math. Mech.* **17**, 641–648 (1968).

S9. On the projection of a self-adjoint operator, *Bull. Amer. Math. Soc.* **74**, 369–372 (1968).

S10. On perturbations of finite rank, *J. Math. Anal. Appl.* **28**, 625–635 (1969).

S11. Some extensions and applications of the new maximum-minimum theory of eigenvalues, *J. Math. Mech.* **19**, 931–944 (1970).

S12. On two complementary variational characterizations of eigenvalues, *Inequalities II*, pp. 375–387. Academic Press, New York, 1970.
S13. Nonclassical choices in variational principles for eigenvalues, *J. Functional Analysis* **6**, 157–164 (1970).
S14. On Fichera's transformation in the method of intermediate problems, *Rend. Accad. Naz. Lincei*, **48**, 302–305 (1970).
S15. Perturbations and intermediate problems for eigenvalues, *Notices Amer. Math. Soc.* **16**, 413 (1969).

STENGER, W., and WEINSTEIN, A.
SW1. On a remark of I. C. Gohberg and M. G. Krein, *Notices Amer. Math. Soc.* **16**, 317 (1969).

TEMPLE, G.
T1. The theory of Rayleigh's Principle applied to continuous systems, *Proc. Roy. Soc. Ser. A.* **119**, 276 (1928).
T2. The computation of characteristic numbers and characteristic functions, *Proc. London Math. Soc.* **29**, 257–280 (1928).
T3. *Proc. Roy. Soc. Ser. A.* **211**, 204–224 (1952).
T4. An elementary proof of Kato's lemma, *Mathematika* **2**, 39–41 (1955).

TREFFTZ, E.
T7. Über Fehlershatzung bei Berechnung von Eigenwerten, *Math. Ann.* **108**, 595–604 (1933).
T8. Die Bestimmung der Knichlast gedrückter, rechteckiger Platten, *Z. Angew. Math. Mech.* **15**, 339–344 (1935).
T9. Die Bestimmung der Knichlast gedrüchter, rechtechiger Platten, *Z. Angew. Math. Mech.* **16**, 64 (1936).

VAN CASTEREN, J. A. W.
V1. On Multiplicative Perturbations of Self-Adjoint Operators (unpublished). Math. Inst., Katholiche Univ., Nijmegen, Netherlands.

VELTE, W.
V2. Über ein Stabilitätskriterium der Hydrodynamik, *Arch. Rational Mech. Anal.* **9**, 9–20 (1962).

WEBER, H.
W1. Über die Integration der partiellen Differentialgleichung, *Math. Ann.* **1**, 1–36 (1869).

WEINBERGER, H. F.
W2. An optimum problem in the Weinstein method for eigenvalues, *Pacific J. Math.* **2**, 413–418 (1952).
W3. Error estimation in the Weinstein method for eigenvalues, *Proc. Amer. Math. Soc.* **3**, 643–646 (1952).
W4. An extension of the classical Sturm-Liouville theory, *Duke Math. J.* **22**, 1–14 (1955).
W5. A Theory of Lower Bounds for Eigenvalues. Tech. Note BN-183, IFDAM, Univ. of Maryland, College Park, Maryland, 1959.

WEINSTEIN, A.
W6. On a minimal problem in the theory of elasticity, *J. London Math. Soc.* **10**, 184–192 (1935).
W7. Sur la stabilité de plaques encastrées, *Compt. Rend.* **200**, 107–109 (1935).

W8. On the symmetries of the solutions of a certain variational problem, *Proc. Cambridge Philos. Soc.* **32**, 96–101 (1936).

W9. Études des spectres des équations aux derivées partielles de la théorie des plaques élastiques, Mémor. Sci. Math. **88** (1937).

W10. Les vibrations et le calcul des variations, *Portugal. Math.* **2**, 36 (1941).

W11. On the decomposition of a Hilbert space by its harmonic subspace, *Amer. J. Math.* **53**, 615–618 (1941).

W12. Separation theorems for the eigenvalues of partial differential equations, *Reissner Anniversary Volume*, pp. 405–416. 1949.

W13. Quantitative methods in Sturm-Liouville theory, *Proc. Symp. Spectral Theory and Differential Problems*, pp. 345–352. Oklahoma State Univ. of Agric. and Applied Science, Stillwater, Oklahoma, 1951.

W14. Variational methods for the approximation and exact computation of eigenvalues, *Symposium for Simultaneous Linear Equations and Determination of Eigenvalues*, Nat. Bur. of Standards, 1953.

W15. Bounds for eigenvalues and the method of intermediate problems, *Partial Differential Equations and Continuum Mechanics*, pp. 39–53. Univ. of Wisconsin Press, Madison, Wisconsin, 1961.

W16. A necessary and sufficient condition in the maximum-minimum theory of eigenvalues, *Studies in Mathematical Analysis and Related Topics*, pp. 429–434. Stanford Univ. Press, Stanford, California, 1962.

W17. On the Sturm-Liouville theory and the eigenvalues of intermediate problems, *Numer. Math.* **5**, 235–245 (1963).

W18. Intermediate problems and the maximum-minimum theory of eigenvalues, *J. Math. Mech.* **12**, 235–246 (1963).

W18a. Intermediate problems and the maximum-minimum theory of eigenvalues (Russian translation of [W18]), *Matematika* **8** (5), 91 (1964).

W19. La teoria di massimo-minimo degli autovalori ed il metodo dei problemi intermedi, *Sem. Ist. Naz. Alta Mat.* pp. 596–609. Edizioni Cremonese, Rome, 1964.

W20. Some results in intermediate problems for eigenvalues, *Atti Simp. Internaz. Appl. An. Fis. Mat.* pp. 218–234. Edizioni Cremonese, Rome, 1964.

W21. Some applications of the new maximum-minimum theory of eigenvalues, *J. Math. Anal. Appl.* **12**, 58–64 (1965).

W22. An invariant formulation of the new maximum-minimum theory of eigenvalues, *J. Math. Mech.* **16**, 213–218 (1966).

W23. On the determination of persistent eigenvalues, *Atti Accad. Naz. Lincei Cl. Sci. Fis. Mat. Nat. Rend.* pp. 515–522. Roma, 1966.

W24. Some numerical results in intermediate problems for eigenvalues, *Numerical Solution of Partial Differential Equations*, J. H. Bramble ed., pp. 167–191. Academic Press, New York, 1966.

W25. On the base problem for a compact integral operator, *J. Math. Mech.* **17**, 217–224 (1967).

W26. On the new maximum-minimum theory of eigenvalues, *Inequalities*, Academic Press, New York, 1967.

W27. The buckling of plates and beams by the method of zero Lagrangian multipliers and zero divisors, *Internat. J. Solids Structures* **4**, 579–583 (1968).

W28. A counterexample to a statement of I. C. Gohberg and M. G. Krein, *Notices Amer. Math. Soc.* **16**, 679 (1969).
W29. Some theoretical ramifications of the intermediate problems for eigenvalues, *Analytic Methods in Mathematical Physics*, R. P. Gilbert and R. G. Newton, eds., pp. 313–334. Gordon and Breach, New York, 1970.

WEINSTEIN, A., and CHIEN, W. Z.
WC1. On the vibrations of a clamped plate under tension, *Quart. Appl. Math.* **1** (1943).

WEYL, H.
W30. Über beschränkte quadratische Formen, deren Differenz vollstetig ist, *Rend. Circ. Mat. Palermo* **27**, 373–392 (1909).
W31. Das asymptotische Verteilungsgesetz der Eigenwerte linearer partieller Differentialgleichungen (mit einer Anwendung auf die Theorie der Hohlraumstrahlung), *Math. Ann.* **71**, 441–479 (1911).
W32. Ramifications, old and new, of the eigenvalue problem, *Bull. Amer. Math. Soc.* **46**, 115–139 (1950).
W33. The method of orthogonal projection in potential theory, *Duke Math. J.* **7**, 411–444 (1940).

ZAREMBA, S.
Z1. L'équation biharmonique et une classe remarquable de fonctions fondamentales harmoniques, *Bull. Acad. Sci. Cracovie.* 147–196 (1907).
Z2. Sur le calcul numérique des fonctions demandées dans le problème de Dirichlet et le problème hydrodynamique, *Bull. Internat. Acad. Sci. Cracovie.* 125 (1909).
Z3. Le problème biharmonique restreint, *Ann. Sci. Ecole Norm. Sup.* **26**, 337 (1909).

ZISLIN, G. M.
Z4. Discussion of the Schrödinger operator spectrum (Russian), *Trudy Moskov. Mat. Obšč.* **9**, 82–120 (1960).

ZISLIN, G. M., and SIGALOV, A. G.
ZS1. The spectrum of the energy operator for atoms with fixed nuclei on subspaces corresponding to irreducible representations of the group of permutations (Russian), *Izv. Akad. Nauk Armjan SSSR Ser. Mat.* **29**, 835–860 (1965).
ZS2. On certain mathematical problems of the theory of atomic spectra (Russian) *Izv. Akad. Nauk Armjan SSSR Ser. Mat.* **29**, 1261–1272 (1965).

Notation Index

Subject Index